BURLEIGH DODDS SCIENCE: INSTANT INSIGHTS

NUMBER 130

Understanding and tackling pests and diseases of forest trees

burleigh dodds
SCIENCE PUBLISHING

Published by Burleigh Dodds Science Publishing Limited
82 High Street, Sawston, Cambridge CB22 3HJ, UK
www.bdspublishing.com

Burleigh Dodds Science Publishing, 1518 Walnut Street, Suite 900, Philadelphia, PA 19102-3406, USA

First published 2026 by Burleigh Dodds Science Publishing Limited

British Library Cataloguing in Publication Data
A catalogue record for this book is available from the British Library

ISBN 978-1-83545-658-3 (Print)
ISBN 978-1-83545-659-0 (ePub)

DOI: 10.19103/9781835456590

Typeset by Deanta Global Publishing Services, Dublin, Ireland

Contents

Series list

Title	Series number
Sweetpotato	01
Fusarium in cereals	02
Vertical farming in horticulture	03
Nutraceuticals in fruit and vegetables	04
Climate change, insect pests and invasive species	05
Metabolic disorders in dairy cattle	06
Mastitis in dairy cattle	07
Heat stress in dairy cattle	08
African swine fever	09
Pesticide residues in agriculture	10
Fruit losses and waste	11
Improving crop nutrient use efficiency	12
Antibiotics in poultry production	13
Bone health in poultry	14
Feather-pecking in poultry	15
Environmental impact of livestock production	16
Sensor technologies in livestock monitoring	17
Improving piglet welfare	18
Crop biofortification	19
Crop rotations	20
Cover crops	21
Plant growth-promoting rhizobacteria	22
Arbuscular mycorrhizal fungi	23
Nematode pests in agriculture	24
Drought-resistant crops	25
Advances in detecting and forecasting crop pests and diseases	26
Mycotoxin detection and control	27
Mite pests in agriculture	28
Supporting cereal production in sub-Saharan Africa	29
Lameness in dairy cattle	30
Infertility and other reproductive disorders in dairy cattle	31
Alternatives to antibiotics in pig production	32
Integrated crop-livestock systems	33
Genetic modification of crops	34

Acknowledgements

Chapters in this Instant Insight are taken from the following sources:

Chapter 1 - Advances in understanding and managing fungal and other diseases of forest trees

Chapter taken from: Stanturf, J. (ed.), Achieving sustainable management of boreal and temperate forests, Burleigh Dodds Science Publishing, Cambridge, UK, 2019, (ISBN 978 1 78676 292 4; www.bdspublishing.com)

Chapter 2 - Pathogens in tropical forests: diversity and management

Chapter taken from: Blaser, J. and Hardcastle, P. D. (eds.), Achieving sustainable management of tropical forests, Burleigh Dodds Science Publishing, Cambridge, UK, 2020, (ISBN 978 1 78676 248 1; www.bdspublishing.com)

Chapter 3 - Advances in understanding and managing insect pests of forest trees

Chapter taken from: Stanturf, J. (ed.), Achieving sustainable management of boreal and temperate forests, Burleigh Dodds Science Publishing, Cambridge, UK, 2019, (ISBN 978 1 78676 292 4; www.bdspublishing.com)

Chapter 4 - Using Verticillium albo-atrum WCS850 to control Dutch elm disease

Chapter taken from: Köhl, J. and Ravensberg, W. J. (eds.), Microbial bioprotectants for plant disease management, Burleigh Dodds Science Publishing, Cambridge, UK, 2021 (ISBN 978 1 78676 813 1; www.bdspublishing.com)

Chapter 1

Advances in understanding and managing fungal and other diseases of forest trees

Tod Ramsfield, Natural Resources Canada, Canada; and Kathy Lewis, University of Northern British Columbia, Canada

1 Introduction

Tree diseases play several important roles in forest ecosystem functions. For example, decay fungi assist with nutrient cycling, and many tree diseases enhance biodiversity by creating gaps in forest canopies. However, disease can cause serious negative effects in stands managed for the production of timber and non-timber forest products. Diseases can affect all parts of the tree, from the roots, stem, and branches, through to the leaves and reproductive parts, and they can result in decreased growth, defects that reduce the value of timber products, or death.

This chapter provides background information on tree disease, and uses case studies of selected pathogens such as *Dothistroma septosporum* (Dorog.)

http://dx.doi.org/10.19103/AS.2019.0057.20
Published by Burleigh Dodds Science Publishing Limited, 2020.

M. Morelet, the causal agent of the disease Dothistroma needle blight, which has been widely introduced and is now causing extensive damage within its native range due to climate change, and *Armillaria ostoyae* (Romagn.) Herink, causal agent of Armillaria root disease and a significant economic issue for foresters. We will outline management strategies for various diseases and describe how the recent introduction of exotic forest pathogens has affected forest ecosystems as well as how climate change is changing host-pathogen interactions leading to uncertainty over future forest health conditions. Due to the prevalence and importance of conifers, the chapter focuses primarily on diseases of conifers in boreal and temperate forests. The intent is to provide examples of diseases that affect different parts of the host tree to illustrate concepts rather than to provide an exhaustive catalogue of every forest disease that occurs in boreal and temperate forests. A recent publication, the *Compendium of Conifer Diseases*, edited by Hansen et al. (2018) provides detailed information on specific conifer pathogens and is an excellent resource. Other excellent resources include *Infectious Forest Diseases* edited by Gonthier and Nicolotti (2013) and *Forest Health: An Integrated Perspective* edited by Castello and Teale (2011), which are both more comprehensive than we can be in a single chapter.

2 Tree disease

Plant disease is defined as 'any malfunction of host cells and tissues that result from continuous irritation by a pathogenic agent or environmental factor and leads to the development of symptoms' (Agrios, 2005). Biotic diseases are caused by pathogens, typically fungi, although bacteria, viruses, and parasitic plants can also cause tree disease. Host infection and disease expression are moderated by the three-way interaction between the host, the pathogen, and the environment, usually described as the disease triangle (Agrios, 2005). The disease triangle concept can be used to describe disease processes in all types of host and pathogen interactions. For host infection to occur, environmental conditions must be suitable to stimulate spore production by the pathogen, transport, germination, and infection of the tree host when it is susceptible to infection. Host susceptibility can also be regulated by environmental conditions, for example, temperature plays a role in timing of bud break and young foliage is more susceptible to infection by foliar pathogens. The physical environment and local weather conditions thus play a critical regulating role in the pathogen infection cycle. Biotic interactions between forest pathogens and other organisms also affect infection and spread through competition and antagonism. Understanding the three-way interaction between the host, the pathogen, and the environment is the basis for many disease management activities. Through manipulating stand structure, for example, by increasing spacing between trees, humidity within the stand can be reduced because the stand is more open, which makes conditions less favorable for pathogens that require high humidity to prevent desiccation of spores, promote spore germination, and favor the infection process. It is also important to recognize that tree pathogens do not always act in isolation and that trees can be subjected to multiple stressors simultaneously. The concept of the decline disease spiral described by Manion (1991) illustrates how 'predisposing factors' of soil, site, and climate affect the ability of the tree to respond to 'inciting factors' such as insect defoliators, drought, frost damage, or salt damage, and 'contributing factors' such as canker and decay fungi, or bark beetles and wood boring insects. These dynamic interactions illustrate the challenges faced by every tree and increase the complexity of forest disease management.

Diagnosis of specific tree diseases involves observation of the physical structures, or 'signs,' of the pathogen and the response of the host, or 'symptoms,' that result from infection. Signs include fungal structures such as mycelium and reproductive structures such as mushrooms or the shelf-like 'conks' that grow on infected tree stems, or acervuli and perithecia embedded in needle tissue. Symptoms include chlorosis (loss of chlorophyll, which causes the foliage to fade in color), thinning crowns, or resin production on the stem. Abiotic disease can occur when trees are exposed to prolonged sub-optimal environmental conditions, such as drought or nutrient deficiencies, or chemicals such as herbicides or pollutants. These abiotic diseases result in symptom production, such as chlorosis of the foliage, without the associated signs of a forest pathogen.

3 Tree disease effects

Stand management objectives influence the perspective of the effects of disease on forests and individual trees. In areas where non-timber values such as biodiversity and habitat are the main objectives, for example parks and protected areas, but also in forests planned for timber production, tree disease caused by native pathogens has many important roles, including habitat creation, nutrient cycling, and acting as an agent of forest succession and biodiversity (Franklin et al., 1987). However, the introduction of exotic forest pathogens into parks and protected areas has resulted in significant damage to trees and ecosystems. For example, *Cronartium ribicola* J.C. Fisch., the causal agent of white pine blister rust was introduced to North America, and its spread into the Rocky Mountain national parks has decimated whitebark pine (*Pinus albicaulis* Engelm.) (Smith et al., 2008) and limber pine (*Pinus flexilis* E. James) (Smith et al., 2013) in some areas. As keystone species, loss of these pines has had serious consequences on habitat and ecosystem function in these high elevation forests. In addition to causing damage within parks, pathogen populations can increase and expand if they are not managed. This can result in the park becoming a source reservoir from which pathogens can emerge to cause damage in adjacent commercial forests. Another negative effect of tree disease in reserves with significant numbers of visitors is that disease effects can cause unstable hazard trees that may result in injury or death if they fail; therefore, if these trees are in high-traffic areas, they must be removed to prevent injury to visitors (Angwin et al., 2012).

In stands managed for timber production or other forest products, disease can become a concern in the following instances: (1) mortality causes stand density to dip below the final stocking values (Fig. 1), (2) disease is spatially clumped within the stand, resulting in areas of high stocking adjacent to areas of low stocking (Fig. 2), or (3) disease causes a periodic or continuous reduction in growth rate, and consequently a decline in tree and stand volume. This can reduce harvested volume or cause a delay in harvest (Fig. 3), or (4) when disease causes significant structural or other defects (such as stain) to host trees, which affects timber quality or can lead to tree failure (Fig. 4). All of these consequences are detrimental in a forest management context because stand productivity is reduced causing losses in volume harvested or necessitating management interventions to mitigate these effects, adding cost and potential delays to harvest.

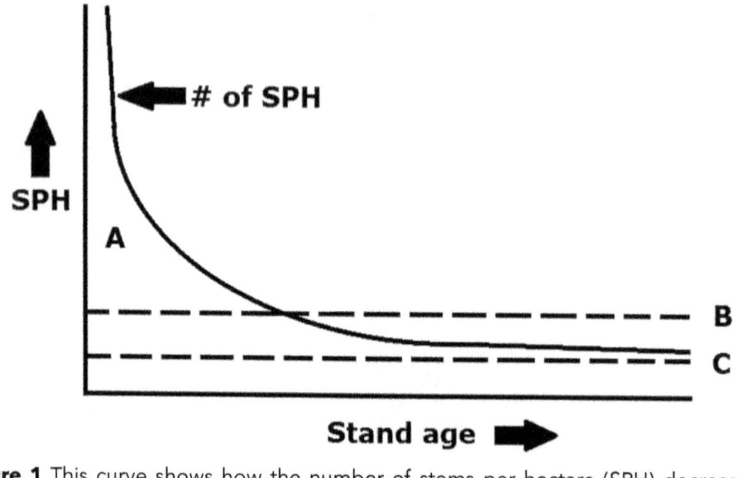

Figure 1 This curve shows how the number of stems per hectare (SPH) decreases as the stand ages. Initially, the number of SPH is high, but stems are rapidly lost due to competition and juvenile mortality. Any mortality that occurs in the region denoted by 'A' is acceptable, as long as disease does not cause the number of SPH to dip below the final stocking level. Dashed line 'B' represents a target final number of SPH; the stand illustrated by the plot has dropped below the final target stocking level and therefore losses to disease are unacceptable. Dashed line 'C' represents an alternative target final stocking level; the number of SPH has not dropped below the target level and any losses to disease that reduce the number of SPH are acceptable.

Published by Burleigh Dodds Science Publishing Limited, 2020.

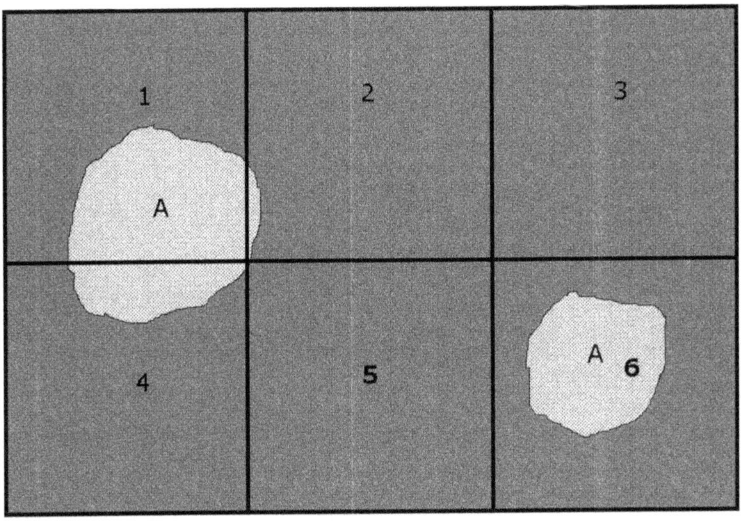

Figure 2 This figure illustrates how the spatial distribution of disease within a stand affects stem density and stand volume. In this scenario, the stand has been divided into six equal sized blocks. A = disease center, caused by *Armillaria ostoyae*. Block 1: Approximately 40% of the stand area is affected by disease. Block 2: Approximately 5% of the stand area is affected by disease. Block 3: The stand is free of disease. Block 4: Approximately 20% of the stand area is affected by disease. Block 5: The stand is free of disease. Block 6: Approximately 30% of the stand area is affected by disease. Each of these blocks is affected by disease to a different degree, with the greatest losses in blocks 1 and 6. Even if the final stocking levels have been achieved in the unaffected portions of blocks 1 and 6, it is likely that the final stocking level has not been achieved for the entire stand and over the long term, volume will be reduced.

4 Classification of tree diseases

Tree diseases can be organized into groups in several ways. They can be organized by the type of agent that causes the disease: fungi, bacteria, nematodes, virus, and abiotic causes. Another approach is to organize them by the mode of action: obligate parasites must have a living host, facultative parasites first kill a host or part of a host and then use the dead tissue for nutrition, and saprophytes utilize dead host material only. A third approach, and the one used in this chapter, is to organize them by the part of the tree the disease affects.

4.1 Foliar disease

Tree pathogens that infect conifer needles can result in reduced canopy productivity as infected needles die and are shed from the tree at a faster rate than normal needle drop that occurs during crown expansion. Foliar pathogens

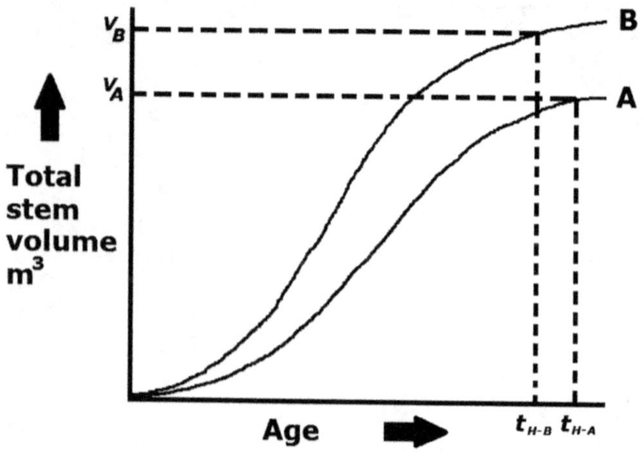

Figure 3 This figure is a simplified diagram to show the effect of non-lethal disease on volume production over time. The figure assumes similar growing conditions and host susceptibility to disease and the only difference between the diseased tree (A) and the non-diseased tree (B) is the presence of a pathogen that is causing a reduction in annual volume production. At time of harvest of tree A (t_{H-A}), the volume accumulated in the stem of tree A (V_A) is lower than the volume accumulated in the stem of tree B (V_B) at time of harvest of tree B (t_{H-B}) due to the presence of a pathogen that has caused disease and associated volume loss. At harvest, diseased tree A has lower final volume and it took longer for growth to plateau than non-diseased tree B that grew to a larger size and was harvested earlier.

can thus reduce the photosynthetic capability of the tree, which is linked to reduced growth and vigor, and with repeated defoliation, can ultimately lead to mortality. If defoliation is not severe enough to cause mortality, annual volume production can be reduced, resulting in an overall reduction in stand productivity. The infection of needles occurs when environmental and host conditions are favorable for the infection process to take place (see Section 2 on the disease triangle), but these conditions do not always occur; therefore, disease severity can vary significantly between years. In years when conditions favor disease development, disease can be widespread and then rapidly disappear from the landscape if conditions do not favor disease development in subsequent years. Thus, the incidence of foliar disease can rise and fall across the landscape. In addition to differences in disease incidence between years, individual trees vary in susceptibility to foliar pathogens due to intraspecific variation based on various resistance mechanisms that are under genetic control (Fraser et al., 2016).

There are many foliar pathogens that can infect conifer needles although the most important diseases identified to date are caused by fungi. In general, foliar fungi are ascomycetes and the spores released from infected needles can either be the sexual ascospores released from asci

Figure 4 Lodgepole pine (*Pinus contorta* var. *latifolia*) with basal infection of *Cronartium harknessii* (J.P. Moore) E. Meinecke (Ex. *Endocronartium harknessii* Hirat.), the causal agent of western gall rust, which has caused the stem to be weakened, leading to breakage at the infection point.

held within a structure such as a perithecium, or asexual conidia released from a pycnidium or an acervulus. These pathogens produce characteristic symptoms, such as banding on the needles, or lesions, but identification of the specific pathogen requires microscopic examination of the pathogen on the needle to make an accurate diagnosis. Two high profile foliar diseases are Dothistroma needle blight, which primarily infects *Pinus* spp., and is caused by the pathogens *D. septosporum* and *D. pini* Hulbary (Barnes et al., 2004), and Swiss needle cast of Douglas-fir, caused by the pathogen *Nothophaeocryptopus gaeumannii* (T. Rohde) (Videira, C. Nakash., U. Braun & Crous) (Videira et al., 2017; Mildrexler et al., 2019). Dothistroma needle blight is explained in more detail in Box 1.

Box 1 Dothistroma needle blight

Dothistroma needle blight (also referred to as red band needle blight) is a foliar disease caused by two closely related pathogens, *Dothistroma septosporum* and *D. pini* (Barnes et al., 2004). The disease can affect multiple *Pinus* spp., and some species of spruce and larch as well as Douglas-fir under optimum conditions. It causes significant damage to pine plantations where either the pathogen or the host, or both, have been introduced (Drenkhan et al., 2016). Recently within its native range in North America it has emerged as a very damaging agent of native pines, likely as a result of climate change (Woods et al., 2005). This disease has caused severe economic damage globally and has been the focus of extensive research (Bradshaw, 2016). Diseased needles usually occur within the lower crown where humidity is higher (Fig. 5), and if the disease persists over multiple years, severe defoliation can result due to repeated casting of dead needles leading to a reduction in growth or even mortality (Bradshaw, 2004). In the fall, small chlorotic lesions appear on infected needles, followed by necrosis, and in the spring, the development of red bands within which black acervuli are formed (Fig. 6) from which conidia are released (Fig. 7). Infected needles are cast over the summer and fall (Parker and Collis, 1966). Environmental conditions are critical for sporulation, and when rainfall and leaf wetness are optimal, conidia are released (Boateng and Lewis, 2015). Multiple resistance mechanisms have been identified (Fraser et al., 2016) and management of this disease has been accomplished with fungicides (southern hemisphere primarily), resistance breeding, and silvicultural practices (Bulman et al., 2016).

Figure 5 Young stand of lodgepole pine (*Pinus contorta* var. *latifolia*) with severe infection by *Dothistroma septosporum*. Note discoloration of the lower crown.

Figure 6 *Dothistroma septosporum* infected needle of lodgepole pine. Note banding pattern and the presence of black acervuli within the band.

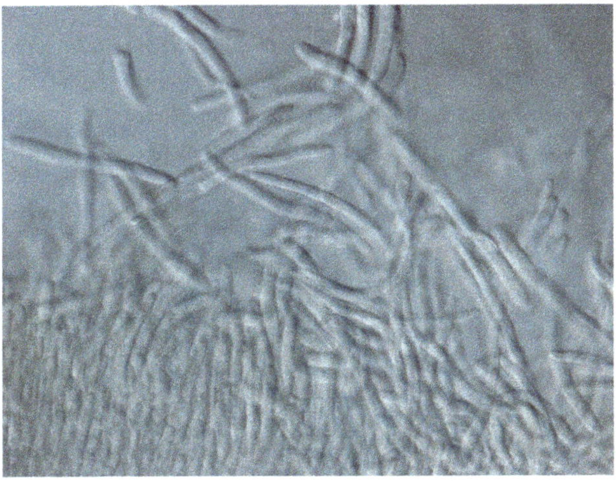

Figure 7 Conidia being released from acervuli of *D. septosporum* on lodgepole pine needle.

4.2 Heart rot of the tree stem

As trees grow, the vascular cambium divides to produce new xylem cells. The newly formed xylem tissue comprises the living sapwood and is responsible for the movement of water and nutrients from the roots to the leaves. New xylem tissue is produced annually, resulting in an increase in the diameter of the tree. As the tree expands in diameter, the cells that comprise the oldest portion of the sapwood die and are converted to heartwood, which is primarily responsible for providing structural support to the stem and serves as storage for metabolic waste products. In a mature tree, the dead interior heartwood is surrounded by a ring of living sapwood. In living trees, the sapwood is highly resistant to decay due to dynamic defense responses that are elicited when the tree is attacked by a pathogen. These include mobilization of resin and other compounds to the wound site.

Although heartwood contains resins, terpenes, and phenolic compounds that increase its resistance to degradation by fungi, some fungi are capable of metabolizing these compounds and can cause decay that reduces the structural integrity of the stem. Some fungi require a wound through the sapwood to bypass the sapwood defenses in order to enter the heartwood. Other fungal species have evolved mechanisms to enter live trees through natural openings, such as branch stubs and even twig scars. In the case of wound-entry heartrot fungi, large scars or other defects can be indicators of the presence of heart rot. In the case of the 'true heart rot' fungi, which enter only through natural openings and only in live trees, there is often little exterior indication of heart rot unless the fungus has produced a fruiting body on the tree stem. Heart rot can lead to stem failure in extreme cases; it can also cause a reduction in wood quality and quantity. Heart rot can occur in both conifers (softwoods) (e.g. *Echinodontium tinctorium* (Ellis & Everh.) Elles & Everh. (Etheridge and Craig, 1976)) and deciduous trees (hardwoods) (e.g. *Phellinus tremulae* (Bond.) Bond. and Boris. (Holmer et al., 1994)); further information regarding *P. tremulae* is outlined in Box 2.

4.3 Vascular wilt diseases of the tree stem

Vascular wilt pathogens live within the vascular tissue and their hyphae grow through xylem vessels, occluding the cells and preventing the movement of water and nutrients. As the tissues distal to the blockage become starved of water and nutrients, they wilt and eventually die, leading to branch flagging, canopy dieback, and eventually tree death. Two serious wilt diseases are Dutch elm disease, a disease of *Ulmus* spp. caused by the pathogens *Ophiostoma ulmi* (Buism.) Nannf., *O. novo-ulmi* Brasier, and *O. himal-ulmi* Brasier and M.D. Mehrotra (Martín et al., 2019), and oak wilt, a disease of *Quercus* spp. caused by the pathogen *Ceratocystis fagacearum* (Bretz) J. Hunt (Juzwik et al., 2011). In conifers, the most important wilt disease is pine wilt disease, caused by the pine wilt nematode (*Bursaphelenchus xylophilus* (Steiner and Buhrer) Nickle (Nematoda: Aphelenchoididae)), which has caused extensive damage as an invasive species in Asia but does not cause a lot of damage within its native range in North America (Tóth, 2011). All three of these agents have a symbiotic association with insects that vector the fungi or nematodes to new host material when the insects feed at the junction of twigs, in order to mature

Box 2 Heart rot of aspen caused by *Phellinus tremulae*

Trembling aspen (*Populus tremuloides* Michx.) of the North American boreal forest, and European aspen (*Populus tremula* L.), are both susceptible to the heart rot agent *Phellinus tremulae* (Bond.) Bond. and Boris. (Holmer et al., 1994; Jones and Ostry, 1998). Comparison of DNA from the internal transcribed spacer (ITS) of isolates of *P. tremulae* from European *P. tremula* and North America *P. tremuloides*, and intersterility between them, indicated that two subgroups of the pathogen exist (Fischer and Binder, 2004). The pathogen can be identified by the presence of large hoof-shaped conks on the stem (Hiratskua et al., 1990) (Fig. 8), although conks are not always present on infected trees. Spread to uninfected trees is via basidiospores, and the infection courts may be branch stubs or other natural openings in the bark, although the specific entry point is unknown. As trees colonized by *P. tremulae* age, the volume of decayed wood also increases as the decay progresses (Wikström and Unestam, 1976). In a study of dead aspen that estimated carbon return to the ecosystem from downed trees, Hogg and Michaelian (2014) found that age and the presence of *P. tremulae* were the best predictors of aspen fall down. Note: *P. tremulae* can only infect living trees; therefore, the pathogen was present in the stem before the aspen fell. This has implications for carbon release as once the stems are on the forest floor, decay proceeds much more quickly than when the trees are still standing because of the higher moisture content of down trees compared to standing trees. Therefore carbon release from the dead tree to the ecosystem, including CO_2 to the atmosphere (Hogg and Michaelian, 2014) proceeds at a much faster rate. Decay columns in living trees have forest management and yield implications (Fig. 9) (Jones and Ostry, 1998), but from a wildlife and biodiversity perspective these large decayed aspen are important because they provide habitat for cavity nesting birds (Aitken and Martin, 2004; Weir et al., 2012), bats (Parsons et al., 2003), and mammals (Aitken and Martin, 2004) (Fig. 10).

their reproductive organs. This leads to establishment of the pathogen and eventual disease. Dutch elm disease and pine wilt disease have both caused extensive damage as introduced pests (Ganley and Bulman, 2016; Osada et al., 2018). It is believed that the Dutch elm disease pathogen was introduced into Europe and North America from Asia (Martín et al., 2019), while the pine wilt nematode is native to North America and was introduced to eastern Asia and western Europe (Tóth, 2011). Elms have been widely planted as amenity trees and Dutch elm disease management programs have been put in place to slow the spread of the disease (Ganley and Bulman, 2016).

Figure 8 Conks of *Phellinus tremulae* on the stem of trembling aspen (*Populus tremuloides*).

4.4 Canker disease of the tree stem and branches

Cankers occur on stems and branches as a result of infection by fungi that cause localized death of the vascular cambium, which results in sunken, discolored tissue. Cankers can persist for multiple years and may cause volume loss or degradation of wood quality. When cankers occur on branches, they often cause branch flagging (dead or dying branches) in the crown. The severity of canker diseases varies due to variation in the virulence of the pathogen and the susceptibility of the host. Annual cankers result from interactions between weak pathogens and vigorous hosts, where the host defense mechanisms are able to surround and isolate the fungus, stopping its growth within a few years. Diffuse cankers result when the pathogen is exceptionally virulent or the host has no evolved defense responses, as in

Figure 9 Trembling aspen (*Populus tremuloides*) infected by *Phellinus tremulae*, resulting in heart rot and decay of the heartwood within the stem. The pathogen causes a reduction in the overall wood quality and weakens the stem, eventually leading to stem failure.

the case of pathogens that are introduced, such as *Cryphonectria parasitica* (Murrill) M.E. Barr, causal agent of chestnut blight. Perennial cankers are in the middle of this continuum, and result from pathogens that are able to overcome host defenses when the host is dormant, but are then surrounded and slowed by host defenses, when the host is actively growing. The result is often a target-like pattern on the stem.

There are several important canker-causing pathogens of conifers, including *Fusarium circinatum* Nirenberg & O'Donnell, the causal agent of pitch canker disease, which may be native to Mexico (Gordon et al., 2015) and has been introduced widely (Wingfield et al., 2008). The pathogen was first observed in southern pines in the United States (Hepting and Roth, 1946). It is now causing significant damage to native stands of *Pinus radiata* D. Don in California (Gordon et al., 2001), and it has been recorded in many countries, including pine plantations in Spain, South Africa, and Chile (Wingfield et al., 2008). Scleroderris canker caused by two races of *Gremmeniella abietina* (Lagerberg) Morelet is present in the boreal forest, and the European race has been introduced into Canada (Hamelin et al., 2000). An important canker disease of *Pinus contorta* var. *latifolia* in Canada is caused by *Atropellis piniphila* (Weir) Lohman and Cash (See Box 3).

Figure 10 *Phellinus tremulae* infected trembling aspen that has been hollowed out by a cavity nesting bird (species unknown).

Box 3 Atropellis canker of lodgepole pine

The canker pathogen *A. piniphila* causes perennial stem and branch cankers on lodgepole pine and occasionally on ponderosa pine (*Pinus ponderosa* Laws.). In their study of the effect of this pathogen on wood quality, Nevill et al. (1989) found that the presence of the pathogen reduced the volume of standard and better grade lumber by up to 40.1%, resulting in a 33.4% reduction in value of wood products from infected trees. The pathogen induces cankers that can be recognized by copious resin production (Fig. 11) and black staining under the bark (Fig. 12). Apothecia formed within the center of cankered tissue release ascospores, which were proven to be infective through artificial inoculation. Infection can occur through unwounded bark and is most common at stem nodes (Hopkins, 1963). During thinning operations, cankered trees should be removed so that spore production within the stand is lowered, reducing the probability of infection (Stanek et al., 1986). The pathogen is common in Alberta and British Columbia, Canada, as well as in portions of the USA (Nevill et al., 1989); it is not present in the European Union and it is therefore a phytosanitary concern and has been subjected to a full pest risk assessment (Jeger et al., 2017).

Figure 11 Atropellis canker of lodgepole pine.

4.5 Root diseases

Root diseases are caused by pathogens that infect and colonize the root tissue of susceptible trees. These diseases lead to reduced capacity for water and nutrient uptake from the soil, resulting in compromised growth, and may cause structural weakness that can result in trees being vulnerable to windthrow. In general, root diseases are widely distributed and cause problems on all continents. Root disease pathogens are often under-represented when assessing pest effects on overall forest productivity as they occur below the soil and they can act over a long period of time (Laflamme, 2010). Major root pathogens include *Armillaria* spp. (Heinzelmann et al., 2019) and *Heterobasidion* spp. (Garbelotto and Gonthier, 2013); there are other root pathogens such as *Onnia tomentosa* (Fr.) P. Karst (Lewis et al., 2004), and *Coniferiporia sulphurascens* (Pilát) L.W.

Figure 12 Lodgepole pine with Atropellis canker, bark removed to show black staining under the bark.

Zhou & Y.C. Dai (Hansen and Goheen, 2000) that also cause significant root disease but they are not as widely distributed. All of these pathogens have broad conifer host ranges. *Armillaria* spp. and *Heterobasidion* spp. are both capable of infecting deciduous species. Tree species vary in their susceptibility to different root disease pathogens based on host resistance mechanisms, although the growing conditions of individual trees also influence susceptibility to disease as site conditions can induce stress. Increased stress can reduce the ability of the tree to resist the infection or colonization process thus leading to increased losses from root disease (Heinzelmann et al., 2019). This is particularly true with *Armillaria* spp. as they are more affected by host vigor than other root disease pathogens. In mature trees, general symptoms of root infection include thinning crowns, reduced growth, and stress cone production (numerous smaller than usual cones). If root disease is present in a stand before harvest, colonized stumps and root systems left behind after logging can serve as an inoculum reservoir from which the pathogen can spread to infect newly planted or naturally regenerated trees (Cleary et al., 2013). The persistence of root disease pathogens on a specific site over multiple rotations, due to the persistence of the pathogen within stumps and root systems, has led to root disease being referred to as a 'disease of the site' (Cleary et al., 2013). An example of a common root pathogen, *A. ostoyae*, is outlined in Box 4.

Once a tree has been killed by *A. ostoyae*, or an infected tree is harvested, the pathogen switches from a parasitic mode where it has to overcome host defenses and is limited in its ability to colonize trees, to a saprophytic mode where it utilizes the roots of the tree for nutrition (Baumgartner et al., 2011). This results in a significant increase in inoculum (inoculum flush), and is one of the

Box 4 Root disease caused by *Armillaria ostoyae*

A. ostoyae is widely distributed in Canada (Laflamme, 2010) and the United States (Hanna et al., 2007), as well as in Europe and Asia (Heinzelmann et al., 2019), although the North American and Eurasian populations are phylogenetically distinct (Klopfenstein et al., 2017). The pathogen has a wide host range, and in western Canada, Douglas-fir and true firs (*Abies* spp.) are the most susceptible to infection while aspen, birch (*Betula* spp.), and poplar are resistant (Morrison and Mallett, 1996). Host infection occurs when the roots of uninfected trees contact the roots of infected trees or when rhizomorphs, black shoelace-like structures that grow through the soil, come into contact with the roots of uninfected trees. Hyphae that emerge from rhizomorphs are able to penetrate cork tissue and then extend through the root tissues, which in some hosts results in a host response that can limit the spread of the pathogen. In susceptible hosts, the cambial tissues of the root are killed, allowing the fungus to spread from the infection point both toward the root collar and away from the infection point toward the root tip. Upon reaching the root collar, the pathogen is able to spread to other primary roots, thus colonizing the root system. When a root is girdled, the distal portion of the root is rapidly colonized by the pathogen, and this may affect the host by causing growth reductions, decay, and eventual death (Morrison et al., 1991). When trees are infected by *A. ostoyae*, the only non-destructive method for identifying the presence of the pathogen is by identifying basidiocarps (mushrooms) at the base of the tree (Fig. 13). The basidiocarps do not persist and they only form when environmental conditions are suitable. Other signs of *A. ostoyae* are mycelial fans under the bark (Fig. 14) and rhizomorphs on root tissue (Fig. 15), although destructive sampling is required to observe these structures.

Figure 13 Basidiocarps of *A. ostoyae* at the base of *A. ostoyae*-infected lodgepole pine.

reasons why Armillaria root disease causes serious damage. Pre-commercial or commercial thinning in stands with Armillaria root disease will increase the rate of spread of the pathogen to residual trees due to this inoculum flush, and similarly spread to planted trees is hastened in clear cut stands compared to rates of spread in undisturbed stands.

5 Parasitic flowering plants

Dwarf mistletoes are obligate parasitic flowering plants of the genus *Arceuthobium* (Viscaceae) that parasitize conifers within the families Pinaceae and Cupressaceae. As they do not photosynthesize, they obtain support, nutrients, and water via the endophytic system that is embedded within host tissues. The lifecycle is initiated when seed produced on female plants are explosively discharged in the fall due to turgor pressure that builds up in the fruit over the summer. Maximum seed dispersal can reach 16 m although 10 m is more typical and most seed are intercepted by foliage within 2-4 m from the

Figure 14 Mycelial fan of *A. ostoyae* under the bark at the base of infected lodgepole pine.

Figure 15 Rhizomorphs of *A. ostoyae* associated with root of lodgepole pine.

Published by Burleigh Dodds Science Publishing Limited, 2020.

source. Seeds intercepted by the foliage of adjacent trees adhere to the needles because of viscin, a sticky coating on the seed. When the seed is wetted by rain, the viscin coating acts as a lubricant, which allows the seed to slide to the base of the needle where it becomes adhered when the viscin dries. Host infection occurs when the radicle that has emerged from the germinated seed is able to penetrate the host branch, initiating the development of the endophytic system in the host's living bark tissues. The dwarf mistletoe plant produces root-like structures that penetrate the host plant cell wall, but not the cell membrane. In that way the host cell is kept alive, and nutrients are transferred to the parasitic plant across the host's semi-permeable cell membrane. In the 2 years following infection, the dwarf mistletoe plant becomes well established in the host branch as the endophytic system is embedded in the branch vascular tissue. As the host branch grows, portions of the parasitic endophytic root system are enveloped by host xylem, and these 'sinkers' help to anchor the parasite in the host. The parasite's aerial system does not emerge until 3 years following infection, and as the genus is dioecious, the plant is either male or female. Pollen is produced on the male flowers, which fertilizes the female flowers, initiating seed development. Seeds are discharged from the fruit in the year following fertilization. It takes a minimum of 6 years for the dwarf mistletoe lifecycle to be completed (Hawksworth and Wiens, 1996).

Some species of *Arceuthobium* induce the formation of witches brooms on the host conifer, by causing the host to make many small branches which benefits the parasite. Brooms alter the crown structure of the infected host and can serve as ladder fuel to bridge ground fires into the canopy within infected stands. Dwarf mistletoes cause volume loss and some can lead to conifer mortality in some host species (Hawksworth and Wiens, 1996). The management of dwarf mistletoes occurs during silvicultural operations; when clearcutting, dwarf mistletoe infected trees should not be left at the edge of the stand as seeds can be cast onto the regenerating conifers. If harvesting is conducted via a variable retention system, it is important that dwarf mistletoe infected trees are removed; otherwise seed from the overstory will be cast onto the regenerating trees in the understory, resulting in infection of the trees in the next rotation (Muir and Geils, 2002).

6 Rust diseases

Rust diseases are caused by basidiomycete fungi in the order Uredinales. These pathogens are obligate parasites that typically have complicated lifecycles that require two hosts and up to five spore stages to complete their lifecycle. For example, *Cronartium comandrae* Peck, the causal agent of comandra blister rust, infects *Pinus* spp. ultimately causing elongated stem cankers, and the alternate hosts *Comandra umbellata* (L.) Nutt. (bastard toadflax) or *Geocaulon lividum* (false toadflax) causing leaf spots. Briefly, aeciospores produced on pine are released in the spring and early summer and infect the leaves of the alternate hosts, which then produce urediniospores that infect the same host, thus increasing the number of infected host plants and the intensity of infection on each plant. Telial columns and teliospores, which then differentiate into basidiospores, are produced on the infected leaves and the basidiospores are released to infect the pine host (Ziller, 1974). The establishment of a rust disease therefore requires the presence of the alternate host in order for the pathogen to become locally established. The invasive rust *C. ribicola*, which infects 5-needle pines and is one of the most damaging pathogens introduced into North America was able to spread because the alternate host (*Ribes* spp.) was widely distributed. Extensive campaigns to remove *Ribes* from the landscape to limit spread of the pathogen to valuable forests were initiated by the United States Department of Agriculture in 1917 although this proved ineffective and the program did not last long (Geils et al., 2010). Rust pathogens in the genera *Cronartium* and *Chrysomyxa* are very important as they infect pines and spruces, respectively. Other important tree rust pathogens reside within the genus *Melampsora* and the pathogens alternate between conifers and *Populus* spp. Not all rust pathogens require an alternate host to complete their lifecycle; *Cronartium harknessii* (J.P. Moore) E. Meinecke, the causal agent of western gall rust is very widespread in western North America and is discussed more fully in Box 5.

The disease is readily identified by the presence of globose galls on branches, which form 2–3 years after infection by spores in the spring, or galls on the stem of the tree that arise from infection of the leader or branch galls that are so close to the stem that they become embedded as the tree grows (van der Kamp et al., 1995). Branch galls are an important source of spores

Box 5 Western gall rust

Cronartium harknessii, the causal agent of western gall rust, was described by Ziller (1974) as 'the most common, most conspicuous, and most destructive stem rust of hard pines in western Canada.' It is widely distributed on 2- and 3-needle pines, including *P. contorta* var. *latifolia*, *Pinus ponderosa* and *Pinus banksiana* Lamb., throughout North America. The pathogen differs from its close relatives in the genus *Cronartium* as it is autoecious and therefore able to infect pines directly without requiring an alternate host. The difference in the lifecycle and the interpretation of nuclear behavior in the germ tube of the germinated spore led Hiratsuka (1969) to erect the genus *Endocronartium* for this pathogen. Recently, the taxonomy has been clarified (Aime et al., 2018) and *C. harknessii* is the current name for the pathogen, which has been referred to in the past as *Endocronartium harknessii* (J.P. Moore) Y. Hirats. and *Peridermium harknessii* J.P. Moore.

Figure 16 Branch of jack pine (*Pinus banksiana*) infected by *Cronartium harknessii*.

for spread of the pathogen to nearby trees, and likely cause a decrease in volume production as nutrients are utilized for gall growth (Fig. 16). Stem galls are important because they lead to stem weakness and breakage (Fig. 4), in addition to reduced wood quality and merchantable volume. The pathogen is closely related to *Cronartium quercuum* f.sp. *banksianae* Burds. & G.A. Snow, a heteroecious rust that requires an alternate host. It is hypothesized that *C. harknessii* is either an autoecious derivative of *C. quercuum* f.sp. *banksianae* or they share a recent common ancestor (Vogler and Bruns, 1998). Since *C. harknessii* does not require an alternate host, it has been recognized as a threat to exotic plantations in countries that have significant investment in *Pinus radiata* D. Don, a known host of *C. harknessii* within the native range of *P. radiata* in California (Ramsfield and Vogler, 2010).

7 Tree diseases caused by species of *Phytophthora*

The group of pathogens within the genus *Phytophthora* (which translated from Greek means 'the plant destroyer') causes disease on all parts of trees and there has been a dramatic increase in the number of forest diseases worldwide caused by *Phytophthora* spp. Historically, *Phytophthora* spp. have caused some of the worst plant diseases in human history, such as *P. infestans* (Mont.) de Bary, causal agent of potato late blight, the disease that led to the Irish potato famine (Fry, 2008). Morphologically, *Phytophthora* spp. resemble filamentous fungi, but they are classified within the Kingdom Stramenopila, phylum Oomycota, due to their motile spores (animalian characteristic) and cell walls composed of cellulose (plant characteristic). *Phytophthora* spp. are either below-ground pathogens that kill fine roots and cause root and root collar rots and bleeding bark cankers or above-ground pathogens that cause leaf necrosis, shoot blights, fruit rot, and bleeding bark cankers. Diseases caused by *Phytophthora* spp. have been recorded in diverse forest ecosystems all around the world and in greatly increasing numbers. As of 2018, 25 serious forest diseases caused by species of *Phytophthora* have been described, all of which have caused significant damage to forest ecosystems (Jung et al., 2018). These include *Phytophthora ramorum* Werres, De Cock & Man in 't Veld, the causal agent of sudden oak death, which affects tanoaks (*Notholithocarpus densiflorus* (Hook. & Arn.) Manos, Cannon & S.H. Oh and coast live oak (*Quercus agrifolia* Née) in coastal California and Oregon (Rizzo et al., 2005) and sudden larch death (*Larix kaempferi* (Lamb.) Carr.) in the United Kingdom (Brasier and Webber, 2010); *Phytophthora pinifolia* Alv. Durán, Gryzenh. & M.J. Wingf., causal agent of 'Daño Foliar del Pino', a foliar disease of *Pinus radiata* in Chile (Durán et al., 2008); *Phytophthora agathidicida* B.S. Weir, Beever, Pennycook & Bellgard, causal agent of Kauri (*Agathis australis* (D. Don) Loudon) dieback in New Zealand (Bellgard et al., 2016); and *Phytophthora austrocedri* Gresl. & E.M. Hansen, the causal agent of 'Mal del Ciprés' of *Austrocedrus chilensis* (D. Don) Pic. Serm. & Bizzarri trees in Patagonia (Taccari et al., 2019). These examples were selected because they were all described recently, the diseases are affecting either commercially important species (*P. radiata*, *L. kaempferi*) or important species in native ecosystems (*L. densiflorus*, *Q. agrifolia*, *A. australis*, and *A. chilensis*), and the origins of the pathogens are unknown. Species of *Phytophthora* can also have a very broad host range and can cause extensive damage in multiple ecosystems. For example, *Phytophthora cinnamomi* Rands is a soil-borne Phytophthora with a very wide host range that has caused damage in North America, Europe, and Australia (Sena et al., 2018). Damage caused by *Phytophthora* spp. to trees in temperate and boreal ecosystems is not as prevalent, although the *P. alni* Brasier & S.A. Kirk complex causes disease of *Alnus* spp. in Europe (Brasier et al., 2004). Stream baiting, a way of detecting the presence of *Phytophthora* spp. in streams within forest ecosystems, has recovered *Phytophthora* spp. in Alaska revealing that a diverse assemblage of species is present there (Reeser et al., 2011), although surveys of alder mortality suggested that the pathogen is not causing extensive disease on alder in Alaska (Adams et al., 2010). *Phytophthora* spp. have caused significant damage to forests and it is believed that the pathogens have been spread in international trade via the trade in live plants for planting (Jung et al., 2018); thus we must remain vigilant to prevent further introductions and damage from this group of pathogens.

8 Abiotic disease

Tree disease results when continual irritation causes symptoms to develop. Abiotic disease can occur as a result of continual irritation from non-living agents such as extended adverse environmental conditions, exposure to chemicals, or nutrient imbalance. Abiotic disease that arises as a result of adverse environmental conditions are difficult to manage, as it is not possible to manage drought and heat stress or temperature inversions, but abiotic disease caused by acid rain or nutrient deficiencies may be addressed through pollution controls or supplementary nutrients.

Recent drought and heat stress attributed to climate change have severely affected forests throughout the world. Widespread tree mortality has been observed and documented in multiple forest types (e.g. van Mantgem et al., 2009). A potential mechanism for these declines links the physiological responses of heat and drought stress, leading to increased host stress and decreased ability of the tree to respond to biotic agents, ultimately resulting in tree death. There is uncertainty with respect to the exact mechanisms that drive these wide-scale mortality events; however, as global temperatures are predicted to continue to rise as a result of climate change, it is possible that these mortality events will increase. These changes to global forest ecosystems would consequently change forest ecosystems and affect societies that rely on ecosystem services provided by forests (Allen et al., 2010). The widespread aspen decline that caused an observed 45 Mt of dead deciduous trees within a survey area of 11.5 Mha at the prairie–boreal interface in Canada has been attributed to severe drought that occurred during 2001–2002 (Michaelian et al., 2011). The widespread and serious losses to forest ecosystems that have already occurred, and may increase, as a result of climate change should be considered as an important motivator to decrease global greenhouse gas emissions.

Another climate-driven phenomenon that has been observed is 'winter drying,' which can occur when temperatures rapidly rise in the winter, causing stomata on needles to open, leading to dehydration of the needle tissue as water is not available to replenish the needle because water in the ground is frozen. The result is needle death and the observation of red needles in the summer following the event. This phenomenon is occasionally observed at the landscape level on mountain sides and has been called 'red belt'. It is driven by temperature inversion in which cool air is overlaid by warmer air and winds and the winter drying effect occurs along an elevational band. It is rare that this phenomenon results in mortality, but defoliation of affected trees does occur (Henson, 1952).

Another example of widespread forest death with an abiotic cause is 'yellow cedar decline', which has been recorded in coastal Alaska and British Columbia. Several years of careful observation and extensive testing of various biotic agents were required to understand the cause of death of yellow cedar (*Callitropsis nootkatensis* (D.Don) D.P. Little). The investigators concluded that a combination of changing climate, soil moisture, and the shallow-rooted nature of yellow cedar combined to lead to mortality. At the landscape scale, the geographic location of patches of dead trees corresponded with elevational limits that varied with latitude, suggesting a climatic influence that was linked with the presence of insulating snow. Tree decline was also related to aspect; decline occurred further upslope on south-facing slopes than north-facing slopes, which corresponded with snow accumulation. Low elevation areas with poor drainage and soil temperatures below -5°C contained dead yellow cedar with fine root damage. Climate change has resulted in changes in suitability within areas that were once suitable for yellow cedar; the tree is very long lived and the reduction in the insulating layer of snow has caused freeze injury to fine roots leading to tree death across the landscape in zones that are no longer suitable for the tree (Hennon et al., 2012).

In addition to climatic stresses, air pollution has caused significant stress to trees due to direct damage to foliage from smelter emissions (Aznar et al., 2009) and as a result of acidification of soils due to sulfur and nitrogen deposition from acid rain (Johnson et al., 2018). This acidification causes changes in soil chemistry that can release toxic compounds, such as aluminum, or deplete important compounds such as calcium, which can have negative effects on tree productivity. Acid deposition can also alter carbon and nitrogen cycling in the organic portion of the soil. The recognition that acid deposition was negatively affecting ecosystems led to air pollution reduction policies in the USA and Europe, which have reduced the amount of sulfur and nitrogen emissions, although monitoring has indicated that there is a time lag between the reduction of pollutants and changes to soil chemistry (Johnson et al., 2018).

In order to achieve their full potential, trees require various micronutrients. Deficiencies in these micronutrients can result in decreased growth and productivity. For example, boron deficiency results in the loss of apical dominance, leading to a decrease in wood quality and volume production (Lehto et al., 2010). Nutrient deficiencies can result in characteristic crown symptoms, such as chlorosis, depending on which element is deficient (Will, 1978). Nitrogen and phosphorous are important for growth, as are calcium, potassium, zinc, copper, and boron, and understanding the effects of these deficiencies on growth has been very important in plantation forestry. Through understanding the effects of nutrient deficiencies on growth, fertilizers can be used to supplement the nutrients available in soil and thus improve growth (Fox et al., 2007).

9 Disease management

To maximize yield and wood quality in forests managed for timber production, the incidence and severity of forest diseases should be minimized. Current management strategies exploit the relationships illustrated by the disease triangle, which links the host, pathogen, and environment to infection and subsequent disease development. Manipulation of one point on the triangle can reduce the incidence and severity of disease and improve growth and yield. Exclusion of pathogens from forest ecosystems is the most effective mechanism for controlling disease and this is the aim of phytosanitary policies and procedures that have been put in place to reduce the probability of establishment of exotic pathogens. For native hosts and pathogens, pathogen exclusion is usually not feasible due to reservoirs of pathogen populations across the landscape. In this case, the goal is to create conditions that limit pathogen build up and spread, thereby reducing the effects of disease. Strategies to achieve this include tree species selection, breeding for resistance, and silvicultural systems that limit pathogen establishment and spread. The following sections describe strategies for forest pathogen management and should be viewed as an introduction to the concepts; all stands are different and a stand-specific prescription will have to consider multiple variables, including local regulations that are beyond the scope of this chapter.

9.1 Tree species selection

Knowledge of pathogen host ranges is fundamental for disease management. Inoculation studies and observations of pathogen incidence have provided important information on host-pathogen interactions that can be used for forest planning. For example, synthesis of the research on Dothistroma needle blight by Drenkhan et al. (2016) has revealed that the pathogens *D. septosporum* and *D. pini* are able to infect 109 host species within the family Pinaceae in the genera *Abies*, *Cedrus*, *Larix*, *Picea*, *Pinus*, and *Pseudotsuga*. Of the 109 species, 20 are highly susceptible, 20 are moderately susceptible, 33 are slightly susceptible, 18 have unknown susceptibility, and 18 have conflicting susceptibility classifications. Similar studies have shown that tree species vary in their susceptibility to *A. ostoyae* (Morrison and Mallett, 1996) and dwarf mistletoes (Hawksworth and Wiens, 1996). These host range studies provide important information for reforestation plans that need to reconcile ecological suitability, management objectives, and future risk of pathogen infection. If the stand management objectives can be met with a tree species that is not susceptible to an abundant pathogen, replanting with that species will reduce the overall losses in the stand.

9.2 Silvicultural techniques

Silvicultural techniques can reduce the incidence and severity of disease through alteration of stand attributes that make conditions unfavorable for disease development. Silvicultural prescriptions can be applied at harvest, during stand establishment, and during thinning operations, as demonstrated in the following examples.

When pathogens are present in an area before harvest, efforts must be taken to avoid spread to planted or naturally regenerating trees. Existing dwarf mistletoe infection can be reduced in regenerating stands by designing clear cuts with cut block boundaries along natural breaks or edges through uninfected areas. This will reduce the spread of seed into the regenerating stand. Partial harvest techniques that leave trees with dwarf mistletoe infection in the overstorey scattered through the stand should be avoided to prevent infection of the regenerating conifers in the understorey (Muir and Geils, 2002).

If root disease is present in the stand before harvest, removal of stumps during or soon after harvest will prevent infection of young trees by root rot agents, such as *Armillaria* spp., that are decaying wood in roots and can spread through root contacts (Cleary et al., 2013). In forests where *Heterobasidion* spp. are present, harvesting can be timed to avoid stump infection by the airborne spores that initiate infection on the surface of the cut stump, a unique feature of this particular root disease pathogen (Gonthier, 2019). Less susceptible regeneration species can also be favored to reduce the probability of root disease or dwarf mistletoe establishment.

In limited cases, control agents can be used to reduce pathogen abundance. In stands where *Heterobasidion* spp. are a risk, the application of urea or borax to freshly cut stumps has been shown to reduce infection of the stump by aerial spores (Gonthier, 2019). Biological control of *Heterobasidion* spp. has also been accomplished through the application of *Phlebiopsis gigantea* (Fr.) Jülich, a fungus that rapidly colonizes the stump, thus preventing infection by *Heterobasidion* spp. (Annesi et al., 2005). The application of chemical control agents has been used extensively in the southern hemisphere to control Dothistroma needle blight; however, chemical control agents are often non-specific and can cause other environmental problems. They are also regulated and not always available for application (Bulman et al., 2016).

In the years following stand establishment, density reduction treatments may be used to improve growing conditions and reduce the number of stems to the prescribed final stocking level. Thinning opens the canopy and reduces humidity and the leaf wetness period, thereby making conditions less favorable for foliar pathogens to sporulate and infect new tissues (Bulman et al., 2016). Thinning operations can also preferentially remove trees with stem infections by pathogens such as *A. piniphila*, reducing disease incidence and removing poor quality trees (Stanek et al., 1986).

When stand management objectives are clear and the potential disease threats are understood, silviculture can be used to maximize growth and yield while minimizing the effects of tree diseases. Silviculture works by manipulating host susceptibility and environmental conditions to induce conditions that are not favorable for disease development.

9.3 Breeding for resistance

Tree improvement programs aimed at improving tree form and growth have been developed and implemented worldwide. Breeding for resistance to forest pathogens has also been a research focus in North America for introduced pathogens such as *C. ribicola* (Sniezko et al., 2014) and *C. parasitica* (Clark et al., 2019) that have caused widespread damage to native forests, and the native pathogen *Cronartium quercuum* (Berk.) Miyabe ex Shirai f. sp. *fusiforme*, which is a major pathogen of southern pines (Sniezko et al., 2014). Surveys of affected forests have identified apparently resistant individuals that have formed the basis of tree breeding programs. Seed from crosses of putatively resistant parents has been screened for resistance through artificial inoculation and field trials. Unlike agricultural crops that have short generation times that can be exploited for rapid genetic improvement, forest trees are long-lived individuals with lengthy generation times, and as a result, these breeding programs represent significant long-term investments. Genomic tools and marker-assisted selection techniques provide promise for rapidly identifying putatively resistant individuals and shortening the breeding cycle. Resistance mechanisms can be qualitative or quantitative. Qualitative resistance (also referred to as 'complete' or 'major gene resistance') involves a single gene, or very few genes, that enable an active response to pathogen infection, preventing pathogen establishment. Quantitative resistance (also referred to as 'incomplete' or 'minor gene resistance') results from the action of multiple genes that make it difficult for the pathogen to become established and tend to be phenotypic traits. Generally quantitative (multigenic) resistance mechanisms are considered more durable even though a lower level of resistance is conferred, because pathogens have a short generation time and there is a risk that resistance will be overcome due to selection pressure on the pathogen. It only takes one virulence gene in the pathogen to overcome a single resistance gene in the host (Sniezko and Koch, 2017).

10 Invasive forest pathogens

Catastrophic damage to forest ecosystems has occurred within the past 150 years due to the establishment of exotic forest pathogens in native forests. Once dominant tree species, such as American chestnut (*Castanea dentata* (Marsh.) Borkh.), have been reduced to minor components of the ecosystem (Rigling and Prospero, 2018), and some native trees, such as whitebark pine, are now threatened with extinction in part due to the presence of exotic pathogens (Shepherd et al., 2018). In order for an exotic organism to become established, the organism must arrive in an area outside of its natural range, conditions must be suitable for establishment, that is, host trees are present and the environmental conditions are suitable, and conditions must be suitable for spread from the point of establishment (Liebhold et al., 1995). These are general requirements for successful establishment of invasive organisms; identification of more specific pathogen traits related to invasiveness has proven difficult (Santini et al., 2013).

Within native communities, coevolution between hosts and pathogens over millennia has limited damage from disease because trees have developed defense responses to pathogens. This dynamic can be disrupted when an introduced pathogen becomes established within areas where the trees have not developed resistance responses and the pathogen is no longer restricted by natural enemies (Brasier, 2008). In these cases, exotic pathogen populations can quickly build and cause severe damage to native hosts. Indeed, Brasier (2008) has described the establishment of an exotic forest pathogen within an expanded range as 'an uncontrolled, potentially dangerous, open-ended experiment in evolution.' Unfortunately, there are several examples of invasive pathogen establishment and subsequent damage to native trees in temperate and boreal forest ecosystems as outlined in Table 1.

Commercial plantation forests established with non-native tree species in the southern hemisphere have also experienced severe damage from exotic invasive insects and pathogens (Wingfield et al., 2015). Awareness of the damage from exotic forest pathogens on native forest ecosystems and planted commercial forests has been met with efforts to prevent further introductions. Global efforts to prevent the incidental movement of pathogens have resulted in the World Trade Organization's Agreement on Sanitary and Phytosanitary Measures and the International Plant Protection Convention (IPPC), which have resulted in a series of International Standards for Phytosanitary Measures (ISPM). These measures provide guidance and regulations to reduce the risk of pathogen establishment (Vettraino et al., 2018; Webber, 2010). Individual countries have established National Plant Protection Organizations (NPPO) that abide by these international agreements that regulate international trade. It is the NPPOs of exporting countries that are responsible for the issuance of phytosanitary certificates that guarantee that export commodities fulfill the phytosanitary requirements of the importing country (Vettraino et al., 2018). The regulatory framework recognizes the importance of reducing the incidental movement of exotic pathogens in trade and has been put in place to reduce the probability of further introductions.

The ISPM standards are internationally agreed upon rules that have been developed to reduce the probability of exotic pest establishment. For example, ISPM 11 provides information on how to conduct a pest risk assessment (PRA) and ISPM 15 focuses on reducing the movement of pests associated with solid wood packaging by establishing treatment regulations. The PRA process

Table 1 Major invasive forest pathogens in temperate and boreal forests

Pathogen	Disease	Primary hosts	Introduced range	Potential native range	Citation
Bursaphelenchus xylophilus	Pine wilt disease	*Pinus* spp.	Asia, Europe	North America	Tóth (2011)
Cryphonectria parasitica	Chestnut blight	*Castanea dentata*, *Castanea sativa* Mill.	USA, Europe	Asia	Rigling and Prospero (2018)
Cronartium ribicola	White pine blister rust	5-needle pines	North America	Asia	Geils et al. (2010)
Dothistroma septosporum and *D. pini*	Dothistroma needle blight	*Pinus* spp.	Europe, Southern hemisphere *Pinus* plantations	Unknown, possibly Canada and Europe	Drenkhan et al. (2016)
Fusarium circinatum	Pitch canker	*Pinus* spp.	Europe, South east and Western USA, numerous countries with exotic *Pinus* plantations	Mexico	Wingfield et al. (2008)
Gremmeniella abietina var. *abietina*	Scleroderris canker, European race	*Pinus* spp.	Eastern North America	Europe	Sanderson et al. (2012)
Hymenoscyphus fraxineus (T. Kowalski) Baral, Queloz & Hosoya	Ash dieback	*Fraxinus excelsior* L.	Europe	Asia	Orton et al. (2018)
Ophiostoma ulmi, *O. novo-ulmi*, and *O. himal-ulmi*	Dutch elm disease	*Ulmus* spp.	North America, Europe	Asia	Martín et al. (2019)
Phytophthora lateralis Tucker & Milbrath	Port-Orford-cedar decline	*Chamaecyparis lawsoniana* (A. Murray) Parl.	Western North America, Europe	Asia	Jung et al. (2018)
Phytophthora ramorum	Sudden oak death	*Notholithocarpus densiflorus*, *Quercus agrifolia*	Western USA, Europe	Asia	Jung et al. (2018)

includes the identification of pathways of pathogen spread. One pathway that has been identified as particularly risky is the plants-for-planting pathway in which live plants are moved around the world in trade (Webber, 2010). By targeting pathways, such as solid wood packaging material or plants-for-planting, a 'manage once, remove many' approach can effectively mitigate the introduction of multiple exotic pests, including pests that may not be considered because they have yet to be identified as potentially serious pests (Evans, 2010).

Damage from the accidental movement of forest pathogens has led countries to maintain diagnostic capacity so that they are equipped to identify any exotic pathogens that may arrive and to continuously improve on diagnostic techniques through research (Bérubé and Nicolas, 2015; Tremblay et al., 2018). Globalization has led to increased trade, travel, and tourism and the threat to native ecosystems from exotic pathogens has increased. Although international regulatory policies have been developed and agencies are in place to enforce these policies, trading nations must continue to assess the risks associated with invasive pathogens and maintain vigilance to prevent the introduction of exotic pathogens. When considered in the context of a changing climate, policy makers and regulators will continue to be challenged due to uncertainty associated with pathogen behavior and potential changes in pathogen risk profiles (Ramsfield et al., 2016).

11 Forest pathogens and climate change

The fifth assessment report of the Intergovernmental Panel on Climate Change outlines recent changes in Earth's atmospheric conditions resulting from greenhouse gas emissions and the consequent increase in global temperatures. Climate change has affected many regions and has resulted in changes to precipitation patterns, snow and ice melt, hydrological systems and geographic ranges of terrestrial, freshwater, and marine species (IPCC, 2014). Future predicted changes to temperature and moisture regimes are expected to result in changes to boreal (Gauthier et al., 2015) and temperate (Millar and Stephenson, 2015) ecosystems through droughts, increased fire frequency, and changes in pest dynamics. It is also likely that the geographic range of tree species will shift as growing conditions change (Price et al., 2013).

As tree diseases result from the interaction between the host, the pathogen, and the environment, changes to environmental conditions will lead to changes in the patterns of disease expression (Mildrexler et al., 2019; Price et al., 2013). Recent reviews have provided insight into how host and pathogen interactions might change, which include environmental changes affecting infection events, host stress induced by climate change leading to increased susceptibility to disease, changes in pathogen distribution, and the introduction of invasive species (Sturrock et al., 2011; La Porta et al., 2008). All types of forest diseases are expected to be affected by climate change. Foliar pathogens are directly affected by climate, for example, damage caused by *D. septosporum* in northwestern British Columbia has already increased, likely due to increased summer rainfall since the 1990s (Sturrock et al., 2011; Woods et al., 2016). A similar situation might exist in coastal Oregon where the incidence of Swiss needle cast of Douglas-fir has also increased recently (Mildrexler et al., 2019). Other pathogens are indirectly affected by climate because they are able to establish in hosts when the hosts are stressed by environmental factors and unable to mount an adequate resistance response. Examples include root disease, cankers, and the pine wood nematode (Sturrock et al., 2011).

As illustrated by the disease triangle, the development of tree disease requires that a combination of susceptible host, virulent pathogen, and suitable environmental factors exist for disease to occur. Future climatic conditions are likely to alter the current distribution of disease on the landscape. Additionally, the adaptive capacity of trees to respond to changes in climate is limited due to long generation times, while pathogens might be able to adapt to a changing climate much more quickly due to their faster reproductive rate. Tree responses to climate change could include changes in their geographic distribution due to changes in water availability and temperature. These are complex interactions, and there is a degree of uncertainty that makes it difficult to predict the future effects of tree diseases on forested landscapes. Given this uncertainty, Sturrock et al. (2011) have highlighted the importance of preparation and planning for anticipating the influence of future climate on forested ecosystems in order to maintain forested ecosystems. They lay out a pathway that includes monitoring, forecasting, planning, and implementing mitigation strategies to ensure the persistence of treed ecosystems on the landscape while environmental conditions are changing. Repercussions from climate change have already been observed in natural and human systems on all continents (IPCC, 2014); managing tree disease in the future will be difficult as changes in host and pathogen interactions and ranges will challenge future forest managers.

12 Conclusion

Forest disease occurs when susceptible hosts are infected by pathogens under environmental conditions that enable the establishment of the pathogen on the host. All parts of trees can be affected by pathogens, and disease can result in decreased volume production, degraded wood products, or death. Disease management strategies can sometimes be used to control the disease and reduce losses in stands managed for timber production. In stands managed for biodiversity and conservation, natural or historic levels of disease can create habitat and promote nutrient cycling; therefore, perspectives on what constitutes a healthy forest depend on management objectives set out in a management plan. There are many threats to native forests, including invasive forest pathogens that have been accidentally introduced, as well as climate change, which has already been cited as a factor that has disrupted the balance between native hosts and pathogens. There is increasing interest and commitment from forest managers to mitigate the effects of climate change and forest pests by managing forests in ways that align with natural disturbance processes and regimes. This has different names in different places (e.g. nature-based silviculture, ecological forestry) but in all cases it is targeted at managing even commercial forests in ways that enhance diversity and habitat through emulation of natural disturbance regimes. This can have positive effects on forest health overall through diversification, but can also have negative effects due to retention of dead or dying trees and limited active management of specific forest health agents. Forest managers are already considering these changes in planning for future forests, but management plans will need to remain flexible and responsive to changes because there is a degree of uncertainty associated with the conditions to which future forests will be exposed.

13 Where to look for further information

13.1 Books

Castello, J. D. and Teale, S. A. (Eds). 2011. *Forest Health: An Integrated Perspective*. Cambridge University Press, New York.

Gonthier, P. and Nicolotti, G. (Eds). 2013. *Infectious Forest Diseases*. CAB International, UK.

Hansen, E. M., Lewis, K. J. and Chastagner, G. A. (Eds). 2018. *Compendium of Conifer Diseases* (2nd edn.). American Phytopathological Society, USA.

13.2 Field guides

Allen, E. A., Morrison, D. J. and Wallis, G. 1996. *Common Tree Diseases of British Columbia*. Natural Resources Canada, Canadian Forest Service, Pacific Forestry Centre, Victoria, BC.

Bergdahl, A. D. and Hill, A. (tech. cords). 2016. *Diseases of Trees in the Great Plains*. Gen. Tech. Rep. RMRS-GTR-355. U.S. Department of Agriculture, Forest Service, Rocky Mountain Research Station, Fort Collins, CO.

Hiratsuka, Y., Langor, D. W. and Crane, P. E. 2004. *Field Guide to Forest Insects and Diseases of the Prairie Provinces* (2nd edn.). Natural Resources Canada, Canadian Forest Service, Northern Forestry Centre, Edmonton, AB.

Holsten, E., Hennon, P., Trummer, L., Kruse, J., Schultz, M. and Lundquist, J. 2008. *Insects and Diseases of Alaskan Forests*. Publication Number R10-TP-140. U.S. Department of Agriculture, Forest Service, Alaska Region.

Myren, D. T., Laflamme, G., Singh, P., Magasi, L. P. and Lachance, D. (Eds). 1994. *Tree Diseases of Eastern Canada*. Natural Resources Canada, Canadian Forest Service, Headquarters, Science and Sustainable Development Directorate, Ottawa, ON.

Oester, P. T., Shaw, D. C. and Filip, G. M. 2019. *Managing Insects and Diseases of Oregon Conifers*. EM 8980. Oregon State University, Oregon State University Extension Service.

14 References

Adams, G. C., Catal, M. and Trummer, L. 2010. Distribution and severity of alder *Phytophthora* in Alaska. In: Frankel, S. J., Klienjunas, J. T. and Palmieri, K. M. (Eds), *Proceedings of the Sudden Oak Death Fourth Science Symposium*. USDA Forest Service, Pacific Southwest Research Station, PSW-GTR-229, pp. 29-49.

Agrios, G. N. 2005. *Plant Pathology* (5th edn.). Elsevier-Academic Press, San Diego, CA.

Aime, M. C., Castlebury, L. A., Abbasi, M., Begerow, D., Berndt, R., Kirschner, R., Marvanová, L., Ono, Y., Padamsee, M., Scholler, M., Thines, M. and Rossman, A. Y. 2018. Competing sexual and asexual generic names in *Pucciniomycotina* and *Ustilaginomycotina* (*Basidiomycota*) and recommendations for use. *IMA Fungus* 9(1), 75–89. doi:10.5598/imafungus.2018.09.01.06.

Aitken, K. E. H. and Martin, K. 2004. Nest cavity availability and selection in aspen - conifer groves in a grassland landscape. *Can. J. For. Res.* 34(10), 2099–109. doi:10.1139/x04-086.

Allen, C. D., Macalady, A. K., Chenchouni, H., Bachelet, D., McDowell, N., Vennetier, M., Kitzberger, T., Rigling, A., Breshears, D. D., Hogg, E. H., Gonzalez, P., Fensham, R., Zhang, Z., Castro, J., Demidova, N., Lim, J., Allard, G., Running, S. W., Semerci, A. and Cobb, N. 2010. A global review of drought and heat-induced tree mortality reveals emerging climate change risks for forests. *Forest Ecol. Manag.* 259(4), 660–84. doi:10.1016/j.foreco.2009.09.001.

Angwin, P. A., Cluck, D. R., Zambino, P. J., Oblinger, B. W. and Woodruff, W. C. 2012. Hazard tree guidelines for forest service facilities and roads in the Pacific southwest region. Report RO-12-01. United States Department of Agriculture Forest Service, Pacific Southwest Region.

Annesi, T., Curcio, G., D'Amico, L. and Motta, E. 2005. Biological control of *Heterobasidion annosum* on *Pinus pinea* by *Phlebiopsis gigantea*. *For. Path.* 35(2), 127-34. doi:10.1111/j.1439-0329.2004.00394.x.

Aznar, J. C., Richer-Laflèche, M., Paucar-Muñoz, H., Bordeleau, M. and Bégin, Y. 2009. Is tree growth reduction related to direct foliar injuries or soil chemistry modifications? *Chemosphere* 76(10), 1366–71. doi:10.1016/j.chemosphere.2009.06.023.

Barnes, I., Crous, P. W., Wingfield, B. D. and Wingfield, M. J. 2004. Multigene phylogenies reveal that red band needle blight of *Pinus* is caused by two distinct species of *Dothistroma*, *D. septosporum* and *D. pini*. *Stud. Mycol.* 50, 551–66.

Baumgartner, K., Coetzee, M. P. A. and Hoffmeister, D. 2011. Secrets of the subterranean pathosystem of *Armillaria*. *Mol. Plant Pathol.* 12(6), 515-34. doi:10.1111/j.1364-3703.2010.00693.x.

Bellgard, S. E., Pennycook, S. R., Weir, B. S., Ho, W. and Waipara, N. W. 2016. *Phytophthora agathidicida*. *For. Phytophthoras* 6(1). doi:10.5399/osu/fp.5.1.3748.

Bérubé, J. A. and Nicolas, G. G. 2015. Alien fungal species on asymptomatic live woody plant material imported into Canada. *Can. J. Plant Pathol.* 37(1), 67–81. doi:10.1080/07060661.2014.986526.

Boateng, K. and Lewis, K. J. 2015. Spore dispersal by *Dothistroma septosporum* in northwest British Columbia. *Phytopathology* 105(1), 69–79. doi:10.1094/PHYTO-06-13-0175-R.

Bradshaw, R. E. 2004. Dothistroma (red-band) needle blight of pines and the dothistromin toxin: a review. *For. Path.* 34(3), 163–85. doi:10.1111/j.1439-0329.2004.00356.x.

Bradshaw, R. E. 2016. Growth and fruition of a pathology research community: the Dothistroma story. *For. Path.* 46(5), 383–7. doi:10.1111/efp.12259.

Brasier, C. M. 2008. The biosecurity threat to the UK and global environment from international trade in plants. *Plant Pathol.* 57(5), 792–808. doi:10.1111/j.1365-3059.2008.01886.x.

Brasier, C. M. and Webber, J. 2010. Sudden larch death. *Nature* 466(7308), 824–5. doi:10.1038/466824a.

Brasier, C. M., Kirk, S. A., Delcan, J., Cooke, D. E., Jung, T. and Man In't Veld, W. A. 2004. *Phytophthora alni* sp. nov. and its variants: designation of emerging heteroploid hybrid pathogens spreading on *Alnus* trees. *Mycol. Res.* 108(10), 1172–84. doi:10.1017/S0953756204001005.

Bulman, L. S., Bradshaw, R. E., Fraser, S., Martín-García, J., Barnes, I., Musolin, D. L., La Porta, N., Woods, A. J., Diez, J. J., Koltay, A., Drenkhan, R., Ahumada, R., Poljakovic-Pajnik, L., Queloz, V., Piškur, B., Doğmuş-Lehtijärvi, H. T., Chira, D., Tomešová-Haataja, V., Georgieva, M., Jankovský, L., Anselmi, N., Markovskaja, S., Papazova-Anakieva, I., Sotirovski, K., Lazarević, J., Adamčíková, K., Boroń, P., Bragança, H., Vettraino, A. M., Selikhovkin, A. V., Bulgakov, T. S. and Tubby, K. 2016. A worldwide perspective on the management and control of Dothistroma needle blight. *For. Path.* 46(5), 472–88. doi:10.1111/efp.12305.

Castello, J. D. and Teale, S. A. (Eds). 2011. *Forest Health: an Integrated Perspective*. Cambridge University Press, New York, NY.

Clark, S. L., Schlarbaum, S. E., Saxton, A. M. and Baird, R. 2019. Eight-year blight (*Cryphonectria parasitica*) resistance of backcross-generation American chestnuts (*Castanea dentata*) planted in the southeastern United States. *For. Ecol. Manag.* 433, 153–61. doi:10.1016/j.foreco.2018.10.060.

Cleary, M. R., Arhipova, N., Morrison, D. J., Thomsen, I. M., Sturrock, R. N., Vasaitis, R., Gaitnieks, T. and Stenlid, J. 2013. Stump removal to control rood disease in Canada and Scandinavia: a synthesis of results from long-term trials. *For. Ecol. Manag.* 290, 5–14. doi:10.1016/j.foreco.2012.05.040.

Drenkhan, R., Tomešová-Haataja, V., Fraser, S., Bradshaw, R. E., Vahalik, P., Mullett, M. S., Martín-García, J., Bulman, L. S., Wingfield, M. J., Kirisits, T., Cech, T. L., Schmitz, S., Baden, R., Tubby, K., Brown, A., Georgieva, M., Woods, A., Ahumada, R., Jankovský, L., Thomsen, I. M., Adamson, K., Marçais, B., Vuorinen, M., Tsopelas, P., Koltay, A., Halasz, A., La Porta, N., Anselmi, N., Kiesnere, R., Markovskaja, S., Kačergius, A., Papazova-Anakieva, I., Risteski, M., Sotirovski, K., Lazarević, J., Solheim, H., Boroń, P., Bragança, H., Chira, D., Musolin, D. L., Selikhovkin, A. V., Bulgakov, T. S., Keča, N., Karadžić, D., Galovic, V., Pap, P., Markovic, M., Poljakovic Pajnik, L., Vasic, V., Ondrušková, E., Piškur, B., Sadiković, D., Diez, J. J., Solla, A., Millberg, H., Stenlid, J., Angst, A., Queloz, V., Lehtijärvi, A., Doğmuş-Lehtijärvi, H. T., Oskay, F., Davydenko, K., Meshkova, V., Craig, D., Woodward, S. and Barnes, I. 2016. Global geographic distribution and host range of *Dothistroma* species: a comprehensive review. *For. Path.* 46(5), 408–42. doi:10.1111/efp.12290.

Durán, A., Gryzenhout, M., Slippers, B., Ahumada, R., Rotella, A., Flores, F., Wingfield, B. D. and Wingfield, M. J. 2008. *Phytophthora pinifolia* sp. nov. associated with a serious needle disease of *Pinus radiata* in Chile. *Plant Pathol.* 57(4), 715–27. doi:10.1111/J.1365-3059.2008.01893.x.

Etheridge, D. E. and Craig, H. M. 1976. Factors influencing infection and initiation of decay by the Indian paint fungus (*Echinodontium tinctorium*) in western hemlock. *Can. J. For. Res.* 6(3), 299–318. doi:10.1139/x76-041.

Evans, H. F. 2010. Pest risk analysis - organisms or pathways? *N. Z. J. For. Sci.* 40(Suppl.), S35–44.

Fischer, M. and Binder, M. 2004. Species recognition, geographic distribution and host-pathogen relationships: a case study in a group of lignicolous basidiomycetes, *Phellinus* s.l. *Mycologia* 96(4), 799–811. doi:10.1080/15572536.2005.11832927.

Fox, T. R., Allen, H. L., Albaugh, T. J., Rubilar, R. and Carlson, C. A. 2007. Tree nutrition and forest fertilization of pine plantations in the southern United States. *South. J. Appl. For.* 31, 5–11. doi:10.1093/sjaf/31.1.5.

Franklin, J. F., Shugart, H. H. and Harmon, M. E. 1987. Tree death as an ecological process, the causes, consequences, and variability of tree mortality. *BioScience* (37), 550–6. doi:10.2037/1310665.

Fraser, S., Martín-García, J., Perry, A., Kabir, M. S., Owen, T., Solla, A., Brown, A. V., Bulman, L. S., Barnes, I., Hale, M. D., Vasconcelos, M. W., Lewis, K. J., Doğmuş-Lehtijarvi, H. T., Markovskaja, S., Woodward, S. and Bradshaw, R. E. 2016. A review of Pinaceae resistance mechanisms against needle and shoot pathogens with a focus on the *Dothistroma-Pinus* interaction. *For. Path.* 46(5), 453–71. doi:10.1111/efp.12201.

Fry, W. 2008. *Phytophthora infestans*: the plant (and *R* gene) destroyer. *Mol. Plant Pathol.* 9(3), 385–402. doi:10.1111/j.1364-3703.2007.00465.x.

Ganley, R. J. and Bulman, L. S. 2016. Dutch elm disease in New Zealand: impacts from eradication and management programmes. *Plant Pathol.* 65(7), 1047–55. doi:10.1111/ppa.12527.

Garbelotto, M. and Gonthier, P. 2013. Biology, epidemiology, and control of *Heterobasidion* species worldwide. *Annu. Rev. Phytopathol.* 51, 39–59. doi:10.1146/annurev-phyto-082712-102225.

Gauthier, S., Bernier, P., Kuuluvainen, T., Shvidenko, A. Z. and Schepaschenko, D. G. 2015. Boreal forest health and global change. *Science* 349(6250), 819–22. doi:10.1126/science.aaa9092.

Geils, B. W., Hummer, K. E. and Hunt, R. S. 2010. White pines, *Ribes*, and blister rust: a review and synthesis. *For. Path.* 40, 147–85. doi:10.1111/j.1439-0329.2010.00654.x.

Gonthier, P. 2019. Frequency of stump infections by *Heterobasidion annosum* s.l. and benefits from urea treatments vary with tree species and season in European alpine forests. *Forest Ecol. Manag.* 434, 76–86. doi:10.1016/j.foreco.2018.12.011.

Gonthier, P. and Nicolotti, G. (Eds). 2013. *Infectious Forest Diseases*. CAB International, Wallingford, UK.

Gordon, T. R., Storer, A. J. and Wood, D. L. 2001. The pitch canker epidemic in California. *Plant Dis.* 85(11), 1128–39. doi:10.1094/PDIS.2001.85.11.1128.

Gordon, T. R., Swett, C. L. and Wingfield, M. J. 2015. Management of *Fusarium* diseases affecting conifers. *Crop Prot.* 73, 28–39. doi:10.1016/j.cropro.2015.02.018.

Hamelin, R. C., Bourassa, M., Rail, J., Dusabenyagasani, M., Jacobi, V. and Laflamme, G. 2000. PCR detection of *Gremmeniella abietina*, the causal agent of Scleroderris canker of pine. *Mycol. Res.* 104(5), 527–32. doi:10.1017/S0953756299002026.

Hanna, J. W., Klopfenstein, N. B., Kim, M. -S., McDonald, G. I. and Moore, J. A. 2007. Phylogeographic patterns of *Armillaria ostoyae* in the western United States. *For. Path.* 37(3), 192–216. doi:10.1111/j.1439-0329.2007.00497.x.

Hansen, E. M. and Goheen, E. M. 2000. *Phellinus weirii* and other native root pathogens as determinants of forest structure and process in Western North America. *Annu. Rev. Phytopath.* 38, 515–39. doi:10.1146/annurev.phyto.38.1.515.

Hansen, E. M., Lewis, K. J. and Chastagner, G. A. (Eds). 2018. *Compendium of Conifer Diseases* (2nd edn.). American Phytopathological Society, St Paul, MN.

Hawksworth, F. G. and Wiens, D. 1996. Dwarf mistletoes: biology, pathology, and systematics. Agricultural Handbook 709. United States Department of Agriculture, Forest Service, Washington DC.

Heinzelmann, R., Dutech, C., Tsykun, T., Labbé, F., Soularue, J. and Prospero, S. 2019. Latest advances and future perspectives in *Armillaria* research. *Can. J. Plant Pathol.* 41(1), 1–23. doi:10.1080/07060661.2018.1558284.

Hennon, P. E., D'Amore, D. V., Schabert, P. G., Wittwer, D. T. and Shanley, C. S. 2012. Shifting climate, altered niche, and a dynamic conservation strategy for yellow-cedar in the north Pacific coastal rainforest. *BioScience* 62(2), 147–58. doi:10.1525/bio.2012.62.2.8.

Henson, W. R. 1952. Chinook winds and red belt injury to lodgepole pine in the Rocky Mountain parks area of Canada. *For. Chron.* 28(1), 62–4. doi:10.5558/tfc28062-1.

Hepting, G. H. and Roth, E. R. 1946. Pitch canker, a new disease of some Southern Pines. *J. For.* 44, 724–44.

Hiratsuka, Y. 1969. *Endocronartium*, a new genus for autoecious pine stem rusts. *Can. J. Bot.* 47(9), 1493–5. doi:10.1139/b69-212.

Hiratsuka, Y., Gibbard, D. A., Bakowsky, O. and Maier, G. B. 1990. Classification and measurement of aspen decay and stain in Alberta. Report NOR-X-314. Forestry Canada, Northwest Region.

Hogg, E. H. and Michaelian, M. 2014. Factors affecting fall down rates of dead aspen (*Populus tremuloides*) biomass following severe drought in west-central Canada. *Glob. Change Biol.* 21, 1968–79. doi:10.1111/gcb.12805.

Holmer, L., Nitare, L. and Stenlid, J. 1994. Population structure and decay pattern of *Phellinus tremulae* in *Populus tremula* as determined by somatic incompatibility. *Can. J. Bot.* 72(10), 1391–6. doi:10.1139/b94-171.

Hopkins, J. C. 1963. Atropellis canker of lodgepole pine: etiology, symptoms and canker development rates. *Can. J. Bot.* 41(11), 1535–45. doi:10.1139/b63-135.

IPCC. 2014. *Climate Change 2014: Synthesis Report. Contribution of Working Groups I, II and III to the Fifth Assessment Report of the Intergovernmental Panel on Climate Change* (Core Writing Team, Pachauri, R. K. and Meyer, L. A., eds). IPCC, Geneva, Switzerland, 151pp.

Jeger, M., Caffier, D., Candresse, T., Chatzivassiliou, E., Dehnen-Schmutz, K., Gilioli, G., Grégoire, J., Jaques Miret, J. A., MacLeod, A., Navajas Navarro, M., Niere, B., Parnell, S., Potting, R., Rafoss, T., Urek, G., Van Bruggen, A., Van Der Werf, W., West, J., Winter, S., Boberg, J., Porta Puglia, A., Vettraino, A. M., Pautasso, M. and Rossi, V. 2017. Pest risk assessment of *Atropellis* spp. for the EU territory. *EFSA J.* 15(7). doi:10.2903/j.efsa.2017.4877.

Johnson, J., Pannatier, E. G., Carnicelli, S., Cecchini, G., Clarke, N., Cools, N., Hansen, K., Meesenburg, H., Nieminen, T. M., Pihl-Karlsson, G., Titeux, H., Vanguelova, E., Verstraeten, A., Vesterdal, L., Waldner, P. and Jonard, M. 2018. The response of soil solution chemistry in European forests to decreasing acid deposition. *Glob. Change Biol.* 24(8), 3603–19. doi:10.1111/gcb.14156.

Jones, A. C. and Ostry, M. E. 1998. Estimating white trunk rot in aspen stands. *North. J. Appl. For.* 15(1), 33–6. doi:10.1093/njaf/15.1.33.

Jung, T., Pérez-Sierra, A., Durán, A., Horta Jung, M., Balci, Y. and Scanu, B. 2018. Canker and decline diseases caused by soil- and airborne *Phytophthora* species in forests and woodlands. *Persoonia* 40, 182–220. doi:10.3767/persoonia.2018.40.08.

Juzwik, J., Appel, D. N., MacDonald, W. L. and Burks, S. 2011. Challenges and successes in managing oak wilt in the United States. *Plant Dis.* 95(8), 888–900. doi:10.1094/PDIS-12-10-0944.

Klopfenstein, N. B., Stewart, J. E., Ota, Y., Hanna, J. W., Richardson, B. A., Ross-Davis, A. L., Elías-Román, R. D., Korhonen, K., Keča, N., Iturritxa, E., Alvarado-Rosales, D., Solheim, H., Brazee, N. J., Łakomy, P., Cleary, M. R., Hasegawa, E., Kikuchi, T., Garza-Ocañas, F., Tsopelas, P., Rigling, D., Prospero, S., Tsykun, T., Bérubé, J. A., Stefani, F. O., Jafarpour, S., Antonín, V., Tomšovský, M., McDonald, G. I., Woodward, S. and Kim, M. S. 2017. Insights into the phylogeny of northern hemisphere *Armillaria*: neighbor-net and Bayesian analyses of translation elongation factor 1-α gene sequences. *Mycologia* 109(1), 75–91. doi:10.1080/00275514.2017.1286572.

Laflamme, G. 2010. Root diseases in forest ecosystems. *Can. J. Plant Path.* 32(1), 68–76. doi:10.1080/07060661003621779.

La Porta, N., Capretti, P., Thomsen, I. M., Kasanen, R., Hietala, A. M. and Von Weissenberg, K. 2008. Forest pathogens with higher damage potential due to climate change in Europe. *Can. J. Plant Path.* 30(2), 177–95. doi:10.1080/07060661.2008.10540534.

Lehto, T., Ruuhola, T. and Dell, B. 2010. Boron in forest trees and forest ecosystems. *For. Ecol. Manag.* 260(12), 2053–69. doi:10.1016/j.foreco.2010.09.028.

Lewis, K. J., Trummer, L. M. and Thompson, R. D. 2004. Incidence of tomentosus root disease relative to spruce density and slope position in south-central Alaska. *For. Ecol. Manag.* 194(1–3), 159–67. doi:10.1016/j.foreco.2004.02.027.

Liebhold, A. M., MacDonald, W. L., Bergdahl, D. and Mastro, V. C. 1995. Invasion by exotic forest pests: a threat to forest ecosystems. *Forest Sci.* 41 (Suppl. 1), a0001–z0001. doi:10.1093/forestscience/41.s1.a0001.

Manion, P. D. 1991. *Tree Disease Concepts* (2nd edn.). Prentice-Hall, Englewood Cliffs, NJ.

Martín, J. A., Sobrino-Plata, J., Rodríguez-Calcerrada, J., Collada, C. and Gil, L. 2019. Breeding and scientific advances in the fight against Dutch elm disease: will they allow the use of elms in forest restoration? *New Forests* 50(2), 183–215. doi:10.1007/s11056-018-9640-x.

Michaelian, M., Hogg, E. H., Hall, R. J. and Arsenault, E. 2011. Massive mortality of aspen following severe drought along the southern edge of the Canadian boreal forest. *Glob. Change Biol.* 17(6), 2084–94. doi:10.1111/j.1365-2486.2010.02357.x.

Mildrexler, D. J., Shaw, D. C. and Cohen, W. B. 2019. Short-term climate trends and the Swiss needle cast epidemic in Oregon's public and private coastal forestlands. *For. Ecol. Manag.* 432, 501–13. doi:10.1016/j.foreco.2018.09.025.

Millar, C. I. and Stephenson, N. L. 2015. Temperate forest health in an era of emerging megadisturbance. *Science* 349(6250), 823–6. doi:10.1126/science.aaa9933.

Morrison, D. and Mallett, K. 1996. Silvicultural management of *Armillaria* root disease in western Canadian forests. *Can. J. Plant Path.* 18(2), 194–9. doi:10.1080/07060669609500645.

Morrison, D. J., Williams, R. E. and Whitney, R. D. 1991. Infection, disease development, diagnosis and detection. In: Shaw III, G. C. and Kile, G. A. (Eds), *Armillaria Root Disease*. Agricultural Handbook No. 691. United States Department of Agriculture, Forest Service, Washington DC, pp. 62–75.

Muir, J. A. and Geils, B. W. 2002. Management strategies for dwarf mistletoe: silviculture. In: Geils, B. W., Tovar, J. C. and Moody, B. (Eds), *Mistletoes of North American Conifers*. RMRS-GTR-98. United States Department of Agriculture, Forest Service, Washington DC, pp. 83–94.

Nevill, R. J., Borden, J. H. and Merler, H. 1989. Reduced volume, grade and value of lodgepole pine lumber caused by Atropellis canker and Stalactiform blister rust. *For. Chron.* 65(1), 36–41. doi:10.5558/tfc65036-1.

Orton, E. S., Brasier, C. M., Bilham, L. J., Bansal, A., Webber, J. F. and Brown, J. K. M. 2018. Population structure of the ash dieback pathogen *Hymenoscyphus fraxineus*, in relation to its mode of arrival in the UK. *Plant Pathol.* 67(2), 255–64. doi:10.1111/ppa.12762.

Osada, Y., Yamakita, T., Shoda-Kagaya, E., Liebhold, A. M. and Yamanaka, T. 2018. Disentangling the drivers of invasion spread in a vector-borne tree disease. *J. Anim. Ecol.* 87(6), 1512-24. doi:10.1111/1365-2656.12884.

Parker, A. K. and Collis, D. G. 1966. Dothistroma needle blight of pines in British Columbia. *For. Chron.* 42(2), 160–1. doi:10.5558/tfc42160-2.

Parsons, S., Lewis, K. J. and Psyllakis, J. M. 2003. Relationships between roosting habitat of bats and decay of aspen in the sub-boreal forests of British Columbia. *For. Ecol. Manag.* 177(1–3), 559–70. doi:10.1016/S0378-1127(02)00448-6.

Price, D. T., Alfaro, R. I., Brown, K. J., Flannigan, M. D., Flemming, R. A., Hogg, E. H., Girardin, M. P., Lakusta, T., Johnston, M., McKenney, D. W., Pedlar, J. H., Stratton, T., Sturrock, R. N., Thompson, I. D., Trofymow, J. A. and Venier, L. A. 2013. Anticipating the consequences of climate change for Canada's boreal forest ecosystems. *Environ. Rev.* 21(4), 322–65. doi:10.1139/er-2013-0042.

Ramsfield, T. D. and Vogler, D. R. 2010. A DNA-based method for detection of *Peridermium harknessii*, the causal agent of western gall rust. *Austral. Plant Pathol.* 39(3), 247–53. doi:10.1071/AP09058.

Ramsfield, T. D., Bentz, B. J., Faccoli, M., Jactel, H. and Brockerhoff, E. G. 2016. Forest health in a changing world: effects of globalization and climate change on forest insect and pathogen impacts. *Forestry* 89(3), 245–52. doi:10.1093/forestry/cpw018.

Reeser, P. W., Sutton, W., Hansen, E. M., Remigi, P. and Adams, G. C. 2011. *Phytophthora* species in forest streams in Oregon and Alaska. *Mycologia* 103(1), 22–35. doi:10.3852/10-013.

Rigling, D. and Prospero, S. 2018. *Cryphonectria parasitica*, the causal agent of chestnut blight: invasion history, population biology and disease control. *Mol. Plant Pathol.* 19(1), 7–20. doi:10.1111/mpp.12542.

Rizzo, D. M., Garbelotto, M. and Hansen, E. M. 2005. *Phytophthora ramorum*: integrative research and management of an emerging pathogen in California and Oregon forests. *Annu. Rev. Phytopath.* 43, 309–35. doi:10.1146/annurev.phyto.42.040803.140418.

Sanderson, L. A., McLaughlin, J. A. and Antunes, P. M. 2012. The last great forest: a review of the status of invasive species in the North American boreal forest. *Forestry* 85(3), 329–40. doi:10.1093/forestry/cps033.

Santini, A., Ghelardini, L., DePace, C., Desprez-Loustau, M. L., Capretti, P., Chandelier, A., Cech, T., Chira, D., Diamandis, S., Gaitniekis, T., Hantula, J., Holdenrieder, O., Jankovsky, L., Jung, T., Jurc, D., Kirisits, T., Kunca, A., Lygis, V., Malecka, M., Marcais, B., Schmitz, S., Schumacher, J., Solheim, H., Solla, A., Szabò, I., Tsopelas, P., Vannini, A., Vettraino, A. M., Webber, J., Woodward, S. and Stenlid, J. 2013. Biogeographical patterns and determinants of invasion by forest pathogens in Europe. *New Phytol.* 197(1), 238–50. doi:10.1111/j.1469-8137.2012.04364.x.

Sena, K., Crocker, E., Vincelli, P. and Barton, C. 2018. *Phytophthora cinnamomi* as a driver of forest change: implications for conservation and management. *For. Ecol. Manag.* 409, 799–807. doi:10.1016/j.foreco.2017.12.022.

Shepherd, B., Jones, B., Sissons, R., Cochrane, J., Park, J., Smith, C. M. and Stafl, N. 2018. Ten years of monitoring illustrates a cascade of effects of white pine blister rust and focuses whitebark pine restoration in the Canadian Rocky and Columbia Mountains. *Forests* 9(3), 138, doi:10.3390/f9030138.

Smith, C. M., Wilson, B., Rasheed, S., Walker, R. C., Carolin, T. and Shepherd, B. 2008. Whitebark pine and white pine blister rust in the Rocky Mountains of Canada and northern Montana. *Can. J. For. Res.* 38(5), 982–95. doi:10.1139/X07-182.

Smith, C. M., Langor, D. W., Myrholm, C., Weber, J., Gillies, C. and Stuart-Smith, J. 2013. Changes in white pine blister rust infection and mortality in limber pine over time. *Can. J. For. Res.* 43(10), 919–28. doi:10.1139/cjfr-2013-0072.

Sniezko, R. A. and Koch, J. 2017. Breeding trees resistant to insects and diseases: putting theory into application. *Biol. Invas.* 19(11), 3377–400. doi:10.1007/s10530-017-1482-5.

Sniezko, R. A., Smith, J., Liu, J.-J. and Hamelin, R. C. 2014. Genetic resistance to fusiform rust in southern pines and white pine blister rust in white pines – a contrasting tale of two rust pathosystems – current status and future prospects. *Forests* 5(9), 2050–83. doi:10.3390/f5092050.

Stanek, W., Hopkins, J. C. and Simmons, C. S. 1986. Effect of spacing in lodgepole pine stands on incidence of Atropellis canker. *For. Chron.* 62(2), 91–5. doi:10.5558/tfc62091-2.

Sturrock, R. N., Frankel, S. J., Brown, A. V., Hennon, P. E., Kliejunas, J. T., Lewis, K. J., Worrall, J. J. and Woods, A. J. 2011. Climate change and forest diseases. *Plant Pathol.* 60(1), 133–49. doi:10.1111/j.1365-3059.2010.02406.x.

Taccari, L. E., Greslebin, A. G., Salgado, M. E. and Vélez, M. 2019. Two conifer species native to Patagonia threatened by *Phytophthora austrocedri*. *For. Path.* doi:10.111/efp.12496.

Tóth, A. 2011. *Bursaphelenchus xylophilus*, the pinewood nematode: its significance and a historical review. *Acta Biol. Szegediensis* 55, 213–7.

Tremblay, É. D., Duceppe, M. O., Bérubé, J. A., Kimoto, T., Lemieux, C. and Bilodeau, G. J. 2018. Screening for exotic forest pathogens to increase survey capacity using metagenomics. *Phytopathology* 108(12), 1509–21. doi:10.1094/PHYTO-02-18-0028-R.

van der Kamp, B. Jvd, Karlman, M. and Witzell, J. 1995. Relative frequency of bole and branch infection of lodgepole pine by western gall rust. *Can. J. For. Res.* 25(12), 1962–8. doi:10.1139/x95-211.

van Mantgem, P. J., Stephenson, N. L., Byrne, J. C., Daniels, L. D., Franklin, J. F., Fulé, P. Z., Harmon, M. E., Larson, A. J., Smith, J. M., Taylor, A. H. and Veblen, T. T. 2009. Widespread increase of tree mortality rates in the western United States. *Science* 323(5913), 521-4. doi:10.1126/science.1165000.

Vettraino, A. M., Potting, R. and Raposo, R. 2018. EU legislation on forest plant health: an overview with a focus on *Fusarium circinatum*. *Forests* 9(9), 568. doi:10.3390/f9090568.

Videira, S. I. R., Groenewald, J. Z., Nakashima, C., Braun, U., Barreto, R. W., de Wit, P. J. G. M. and Crous, P. W. 2017. *Mycosphaerellaceae* - chaos or clarity? *Stud. Mycol.* 87, 257-421. doi:10.1016/j.simyco.2017.09.003.

Vogler, D. R. and Bruns, T. D. 1998. Phylogenetic relationships among the pine stem rust fungi (*Cronartium* and *Peridermium* spp.). *Mycologia* 90, 244-57. doi:10.1080/00275514.1998.12026904.

Webber, J. 2010. Pest risk analysis and invasion pathways for plant pathogens. *N. Z. J. For. Sci.* 40(Suppl.), S45-56.

Weir, R. D., Phinney, M. and Lofroth, E. C. 2012. Big, sick and rotting: why tree size, damage, and decay are important to fisher reproductive habitat. *For. Ecol. Manag.* 265, 230-40. doi:10.1016/j.foreco.2011.10.043.

Wikström, C. and Unestam, T. 1976. The decay pattern of *Phellinus tremulae* (Bond.) Bond. Et Borisov in *Populus tremulae* L. *Eur. J. Forest Pathol.* 6, 291-301. doi:10.1111/j.1439-0329.1976.tb00539.x.

Will, G. M. 1978. Nutrient deficiencies in *Pinus radiata* in New Zealand. *N. Z. J. For. Sci.* 8, 4-14.

Wingfield, M. J., Hammerbacher, A., Ganley, R. J., Steenkamp, E. T., Gordon, T. R., Wingfield, B. D. and Coutinho, T. A. 2008. Pitch canker caused by *Fusarium circinatum* - a growing threat to pine plantations and forests worldwide. *Austral. Plant Pathol.* 37(4), 319-34. doi:10.1071/AP08036.

Wingfield, M. J., Brockerhoff, E. G., Wingfield, B. D. and Slippers, B. 2015. Planted forest health: the need for a global strategy. *Science* 349(6250), 832-6. doi:10.1126/science.aac6674.

Woods, A., Coates, K. D. and Hamann, A. 2005. Is an unprecedented Dothistroma needle blight epidemic related to climate change? *BioScience* 55(9), 761-9. doi:10.1641/0006-3568(2005)055[0761:IAUDNB]2.0.CO;2.

Woods, A. J., Martín-García, J., Bulman, L., Vasconcelos, M. W., Boberg, J., La Porta, N., Peredo, H., Vergara, G., Ahumada, R., Brown, A. and Diez, J. J. 2016. Dothistroma needle blight, weather and possible climatic triggers for the disease's recent emergence. *For. Path.* 46(5), 443-52. doi:10.1111/efp.12248.

Ziller, W. 1974. The tree rusts of western Canada. Publication No. 1329. Canadian Forest Service, Victoria, BC.

Chapter 2

Pathogens in tropical forests: diversity and management

Steve Woodward, University of Aberdeen, UK; and H. Tuğba Doğmuş, Isparta University of Applied Sciences, Turkey

1 Introduction

In general, natural forests are considered to be balanced ecosystems in which the plants have co-evolved with the native pests and pathogens and, in the absence of major climatic disruption, the damage is limited. Susceptible plants are killed by problems before attaining maturity and, hence, future generations of the forest plants arise from parents with resistance to the range of pests and pathogens present in the ecosystem. Arguably, this process is a major component in the evolution of organisms, enabling only the healthiest offspring to survive. Climatic disturbance may result in the forest environment suddenly becoming suboptimal for tree growth, which will reduce the capacity of the plants to defend against pests and pathogens. Human interventions in forestry, including both management of natural forests and establishing plantations, also often cause suboptimal conditions, making the stock more susceptible to existing pests and pathogens. Further problems have arisen when global

http://dx.doi.org/10.19103/AS.2020.0074.35

trade or other human activities result in inadvertent transport of alien pests and pathogens between continents (Santini et al. 2018), some of which establish on host plants with no co-evolutionary history with the invasive species: these problems often result in serious tree health issues. Sometimes the problems are catastrophic for a given genus of trees.

2 Root diseases

Under natural conditions, it is likely that the fine roots of plants are under continuous attack from a range of soil-borne pathogens of low virulence. We detect these pathogens as problems in specialized situations, such as in tree nurseries, but in the forest and under reasonable conditions, older trees replace parts of the root system continuously. The phenomenon of fine root turnover is well-known in plant biology (Van Noorwijk et al. 2004), and it is certain that fine root pathogens in the genera *Phytophthora*, *Pythium*, *Fusarium* and *Rhizoctonia*, among others, play a role in this process.

Fungal and fungus-like pathogens are believed to play a role in seed dispersal for some tree species preventing con-generic regeneration in close proximity to the mother tree (e.g. Augsperger 1983; Packer and Clay 2000). The mechanism for this process is thought to involve the build-up of the pathogen population in the rhizosphere of a tree as it ages: the mature tree tolerates the presence of the pathogen, but the roots of seedlings are rapidly overwhelmed by the soil inoculum and the young plants die. This process indirectly encourages seed dispersal and healthy growth of seedlings when germination occurs at some distance from the mother tree (Packer and Clay 2000).

Under some circumstances, however, conditions in the forest may become favourable to the pathogen, or unfavourable to the tree, and certain pathogens can cause considerable damage to mature trees. Several species of Oomycota, particularly in the genus *Phytophthora*, can cause serious problems on trees when the environmental conditions favour the pathogens. Such favourability includes periods of soil saturation, interspersed with dry conditions (Hansen et al. 2012). The classic example of these pathogens is *Phytophthora cinnamomi*. Native in the Sulawesi region of Far East Asia (Arenz and Simpson 1986), *P. cinnamomi* was spread globally by human activities and now causes problems not only in tropical and subtropical regions, but also in warm temperate regions (Bergot et al. 2004). Examples of the extreme damage that *P. cinnamomi* can cause are typified by 'Jarrah dieback', affecting *Eucalyptus marginata,* other *Eucalyptus* spp. and a wide range of other plants in Western Australia, Queensland and Victoria (Shearer et al. 2004; Weste 2003). Inadvertent global distribution of *P. cinnamomi* probably began in the early nineteenth century (Santini et al. 2013) and has resulted in severe epidemics in several regions of the world.

Species like *P. cinnamomi* are spread in infested soil, in run-off from contaminated sites, and, initially, on roots of young plants raised in infected nurseries (e.g. Jung et al. 2015). In Western Australia, strict regulations are applied using good hygiene and sanitation of forest machinery, and all human activities in affected areas must apply these protocols to machinery leaving the area (Colquhoun and Hardy 2000).

Other Oomycota may also cause dieback and mortality of tropical trees: for example, *Phytopythium cucurbitacearum* is implicated in dieback of *Cedrela odorata* in Trinidad, causing severe dieback in exotic provenances of the host tree (Woodward, Boa, Belbahri, Belka, Taylor & Siasou et al., unpublished; Fig. 1). The genus *Phytopythium*, recently separated from *Pythium* (clade K; De Cock et al. 2015), is widespread in tropical regions.

Fine roots of trees of all ages are attacked by soil-borne true fungi, including the wide host-range *Macrophomina phaseolina* (Botryosphaeriales; charcoal root rot or black root rot; e.g. Barnard 1994) and the Hyphomycete *Cylindrocladium* spp. (Nectriaceae; teleomorphs are in the genus *Calonectria*; Crous 2002). Where microsclerotia infest soils, very high losses of roots may be incurred, leading to rapid mortality, especially in nursery situations. *M. phaseolina* is associated with high soil temperatures during the growing season (Arias et al. 2013).

2.1 Root decay fungi

Larger, secondarily thickened roots of trees are attacked by a range of generally non-host-specific fungi capable of producing specialized peroxidase enzymes that degrade lignified tissues. These fungi are essential in the mineralization of woody tissues, playing a primary role in nutrient cycling (Boddy and Watkinson 1995). Well-known examples in tropical forests are the Hymenomycetes *Phellinus noxius* (brown root disease (Fig. 2)), *Armillaria* spp. (Armillaria root

Figure 1 *Cedrela odorata* affected by *Phytopythium cucurbitacearum* in Trinidad. (a) symptoms of severe dieback in the crown of a tree; (b) bleeding canker at the stem base, typical of oomycete infections. (Photos courtesy of E. Boa).

Figure 2 The formation of a typical brown, fungal 'sock' around the lower stem of a tree - in this case *Araucaria cunninghamii*, infected by *Phellinus noxius*. (Courtesy of G. Pegg, Queensland Department of Agriculture and Fisheries, Australia).

rot (Fig. 3)), *Rigidoporous microporus* (white root rot), *R. vinctus* (poria root rot) and *Ganoderma* spp. (red root rot; Fig. 4; Lundquist 1987; Mohammed et al. 2014).

These types of decay-causing fungi can usually be regarded as specialized saprotrophs, which predominantly colonize the dead heartwood of trees and cannot invade living tissues, due to the high water content (Rayner and Boddy 1988). Some decay fungi, however, invade living tissues, such as inner bark (secondary phloem), vascular cambium and sapwood, in addition to degrading lignified heartwood. Invasion by species such as *Armillaria* (Fig. 3) is known to require high inoculum potential from established infections in woody substrates in the soil, perhaps dead roots or tree stumps. Spores or mycelium of many such fungi may grow only a short distance through the soil from the infected substrate but, generally, can only invade root tissues which come into direct contact with infected woody debris from the previously colonized tree root systems (i.e. *P. noxius*; most tropical *Armillaria* spp.; root-infecting *Ganoderma*

(a) (b) (c)

Figure 3 *Armillaria* spp. are among the most damaging of root pathogens affecting trees globally. (a) typical fruiting bodies of *Armillaria*; (b) infected trees that have recently died may have a mycelial sheath evident if the bark is removed from the lower trunk; (c) disease centre in a pine plantation.

spp.; Lee 2004; Mohammed et al. 2014). An infected inoculum source, such as an old stump, root debris or fallen stem from the previous woody vegetation, is usually at the centre of these infections, leading roughly circular-expanding disease foci.

Figure 4 Symptoms of *Ganoderma* spp. infection in an Acacia plantation. (a) crowns of the trees are showing symptoms of severe thinning, indicating a root disease problem; (b) *Ganoderma* spp. fruiting bodies on the lower stem of *Acacia* spp. (Courtesy of L. Su See).

Based on observations of the impacts of root decay fungi, it appears that the organisms are widespread in undisturbed indigenous tropical forests, existing as saprotrophs, but occasionally causing mortality to native trees that are otherwise stressed. Windthrow, stem breakage and felling activities upset the ecosystem balance and provide woody substrates for additional colonization by saprotrophic decay fungi. In natural ecosystems, however, it appears that disease foci rarely enlarge rapidly. An extreme example of serious ecosystem disturbance by human activities is the clear-felling of native dipterocarp-dominated forests in Southeast Asia, to replace the trees with oil palm or Acacia plantations: in the plantations, large outbreaks of root diseases occur, leading to serious yield losses in the crops (Tsui et al. 1998).

Soil conditions have significant impacts on the development and spread of root diseases. Moreover, the size of the food base inhabited by fungi attacking secondarily thickened tissues is of importance, providing resources for persistence and penetration of the roots of nearby trees. Stumps produced by felling operations are particularly suitable and are long-term disease sources. In plantations of palms or rubber, for example, stumps of the primary forest, or of the previous crop, are excellent inoculum sources for the planted crop (Ariffin et al. 2000).

3 Stem diseases

Stems of trees are attacked by a range of fungal pathogens, resulting in sunken cankers, swollen galls or internal decay of the wood. Many of the pathogens causing these problems require wounds for initial establishment. Stem decay fungi are particular problems in plantations, where considerable loss of timber can occur.

Cankers and galls weaken stems (and branches), making the trees susceptible to breakage. A particularly damaging example of canker is 'pink disease' caused by *Erythricium salmonicolor*, and present throughout the tropics. This pathogen has a very wide host range, including rubber, and is particularly well-known on *Eucalyptus* spp. in Southern India (Keane et al. 2001); it has also caused problems on many other tropical tree species (Roux and Coetzee 2005). Small lesions develop in young bark, with the superficial white mycelium growing over unaffected bark. In high humidity, the mycelium becomes more visible, turning pink in colour. The underlying phelloderm and secondary phloem tissues become necrotic and contain pink pustules of sterile mycelia. The fruiting body of the pathogen is a pink resupinate basidiome that forms on the under-surface of affected branches, and orange-red pustules of the anamorph also often form on upper surfaces of branches. Death of young branches occurs, leading to extensive dieback symptoms that eventually reach the main stem. On other trees, cankers girdle branches and stems, resulting in top-die back, stem breakage or mortality. Infections are dormant in dry weather, making diagnosis difficult at these times.

Other fungi causing cankers on *Eucalyptus* spp. include *Diaporthe cubensis* (syn. *Cryphonectria cubensis*, *Endothia havanensis*; Boerboom and Maas 1970; Hodges 1980), which led to high rates of mortality in young plantations of *E. grandis* and *E. saligna* in Surinam (Boerboom and Maas 1970), although the pathogen occurs in many parts of the world where *Eucalyptus* spp. are grown (Conradie et al. 1990; Wingfield et al. 2008a). A number of different *Cytospora* spp. also cause stem cankers of variable sizes on *Eucalyptus*.

Several canker-causing rust pathogens are known to affect tropical angiosperm trees; characteristically, rust pathogens have complex life cycles, affecting two (or more) unrelated hosts (macrocyclic rusts). In contrast, microcyclic rusts repeat their life cycles on the same host. Few rust pathogens, however, cause significant stem damage on important tropical forest trees. Examples include stem rust of *Cordia alliodora* caused by *Puccinia cordiae* in Central and South America (e.g. Pardo Cardona 1998) and the Caribbean, and stem rust of *Paraserianthes falcataria* (syn. *Adenanthera falcataria*; *Albizia falcataria*; *Falcataria moluccana*), caused by *Uromycladium tepprianum,* in the Philippines (Old and Santos Cristovao 2003).

Decay fungi, often similar to those attacking secondary roots (Section 2.1), also frequently attack tree stems. It should be emphasized that the decay process is natural and of great importance in forest ecosystems, as it releases nutrients sequestered in the long-lived woody tissues for recycling in the ecosystem. Decay becomes a problem in tropical plantations and other situations where timber is required by humans. Although it was long believed that decay fungi colonize wounds to enter stems and larger branches, more recent work has thrown some doubt on the hypothesis (Parfitt et al. 2010).

Decay fungi are almost all Basidiomycota, in the class Agaricomycetes, although there are a few Ascomycota that cause decay in standing trees. Within the Agaricomycetes, two broad groups are recognized:

1. White rot fungi are the most efficient wood-degrading fungi, producing manganese-dependent lignin peroxidases, and, in most species, laccases. Some species produce a separate enzyme family known as lignin peroxidases. These peroxidases are 'stress' enzymes produced in adverse growing environments and can be induced artificially by increasing oxygen tension above cultures of producing fungi. White rot decay results in bleaching, rendering a pale colour to the wood, hence the common name. Tropical examples include *Stereum sanguinolentum* in *Pinus patula* damaged by elephants (Fig. 5) or water buffalo (Lundquist 1987; Dublin 1995); this fungus is also common in temperate regions. Several species of *Ganoderma* are also problematic in tropical forests.
2. Brown rot fungi: In contrast to white-rot fungi, brown rot fungi are less adapted to degradation of lignin in cell walls, producing enzymes that attack carbohydrate polymers (cellulose, xylans), but not 'ligninases'. The decay is characterized by a dark brown colour and crumbly texture. Despite the lack of manganese or lignin peroxidases, these fungi are still capable of oxidizing the lignin-based phenolic compounds, which produces the brown colour of the decayed tissues. An example affecting tropical trees is *Pseudophaeolus baudonii*.

Most decay fungi are considered highly specialized saprotrophs rather than true pathogens as they attack the heartwood, which in most trees is dead. Despite the presence of many potentially toxic metabolites, decay fungi are adapted to detoxify such compounds. The changes in woody tissues that can often be observed immediately adjacent to decayed areas are, therefore, mainly due to simple responses of the wood to the production of enzymes by the fungus. Decayed wood inside living trees is compartmentalized by changes in the surrounding tissues (e.g. Shigo and Marx 1977), but the compartment walls are often breached by further growth of the decay fungi (Pearce 1996).

Figure 5 The wound-infecting pathogen, *Stereum sanguinolentum,* can cause severe decay in stems of Pinaceae (a). Fruiting bodies are typically khaki-coloured and resupinate (b). (Figure 5b courtesy of J. Viola, Northeastern University, Bugwood.org).

Fungi are limited in their abilities to grow in wet environments. This feature, coupled with the relatively high oxygen tension of xylem sap, is probably responsible for the rarity of sapwood decay in living trees (Boddy and Rayner 1983). As a living tissue, sapwood can also respond to the presence of invading fungi, forming Shigo's wall 4, which is undoubtedly the strongest of the four walls defined in the CODIT hypothesis (Pearce 1996). In stress conditions, however, including drought or waterlogging, fungi may breach wall 4 and enter the functional sapwood.

Fungi in the Ophiostomataceae (Ascomycota) colonize sapwood and heartwood of trees through wounds but are restricted to utilize easily assimilable sugars and amino acids in the host tissues. The problem caused by many of these fungi is timber staining, which reduces market values for the wood (Jacobs and Wingfield 2001). Certain species within the family such as *Ophiostoma novo-ulmi* (cause of Dutch elm disease) and *Leptographium* spp. (black stain root disease) are highly damaging pathogens, however, resulting in mortality of affected trees (Hadfield et al. 1986).

4 Wilt diseases

Vascular tissues, specifically the xylem, may be infected by certain fungi (and bacteria), resulting in wilt. The symptoms of wilt are not caused by the pathogen itself blocking the xylem vessels or tracheids, but by the response of the host plant to the presence of the pathogen. The plant attempts to close-off affected vessels, as might happen if the sapwood is damaged physically, but wilt pathogens produce toxins which spread rapidly in the vascular system and open stomata and induce systemic failure: the plant is no longer able to regulate water relations. A consequence of this problem is that the foliage and young shoots wilt. A notable symptom of wilt infections is the production of polyphenol oxidase enzymes by xylem companion cells, resulting in

Figure 6 Wilt on *Pterocarpus indicus* in the Seychelles is caused by a strain of *Fusarium oxysporum*. (a) severe thinning in the crown of an infected tree; (b) staining in the vascular cambium and sapwood on the main stem. (Courtesy of E. Boa).

dark streaking around the affected vessels. Depending on the host species, companion cells may also produce tyloses or gums/gels into the affected xylem vessels, which are responsible for blockages.

Many wilt pathogens, particularly those in the genera *Verticillium* and *Fusarium*, are soil-borne and enter host plants through fine roots. Other species, including *Ophiostoma novo-ulmi* and *Leptographium* spp. (Section 3), are vectored by insects. Several wilt diseases are well-known in tropical regions, including sissoo wilt on *Dalbergia sissoo* in the Indian subcontinent (Al Adawi et al. 2013), wilt of *Pterocarpus indicus* in the Seychelles (Fig. 6; Boa 2014, personal communication) and other more recently discovered *Ceratocystis* wilts in various regions of the world (Harrington 2013), including *C. fimbriata* in teak (*Tectona grandis*) plantations in Brazil (Firmino et al. 2012). Wilt of *D. sissoo* was for many years considered to be caused by a forma of *Fusarium solani* (Rajput et al. 2011), but more recently, it was suggested to result from infection by *Ceratocystis manginecans* (Al Adawi et al. 2013).

5 Foliage pathogens

Young foliage and shoots (before the onset of suberization) of trees are attacked by many different pathogens, including species with simple life cycles, repeating once, twice or several times each year (e.g. Dothistroma needle blight of pines), or complex life cycles extending over two or more years. The full life cycles of rust pathogens involve at least two alternate, unrelated hosts. Foliage and shoot diseases are common and widespread on trees, although the amount of damage is usually too little to cause notable growth reductions. Occasionally, however, severe disease does occur on forest trees in tropical countries, either following some separate disturbance or on exotic tree species planted in regions favourable to the development of the disease. More recently, these problems have increased on indigenous trees, partly due to human transport of alien invasive pathogens from their region/continent of evolution to naïve regions (e.g. Santini et al. 2013; Santini et al. 2018), and due to climate change.

Dothistroma needle blight (DNB; red band needle blight; Fig. 7), which mainly affects *Pinus* spp., is an example of a pathogen with a simple life cycle. Two species of *Dothistroma*, *D. pini* and *D. septosporum* (Ascomycota) cause the disease, with the abundance of each species varying in different parts of the world. The disease was first described in western Russia in 1911, but in the mid-twentieth century began to attack monocultures of various *Pinus* spp., particularly *P. radiata*, in the uplands of East Africa, in Chile and New Zealand (Bulman et al. 2013; Drenkhan et al. 2016), causing defoliation and mortality. In the periods of high humidity with suitably cool temperatures, frequently encountered in the highlands of East and Central Africa, *D. septosporum* has a short life cycle of approximately 40 days, with a number of generations per year, leading to damage of epidemic proportions within a short time. Conidia are spread by the wind in mist droplets after splash dispersal from acervuli (asexual fruiting bodies), leading to distribution between plantations, and, over longer periods, between regions. It is believed that the near-simultaneous first occurrence of DNB in the highlands of East Africa, Central Africa, Chile and New Zealand was due to infected needles contaminating seed lots moved intercontinentally for planting (Bulman et al. 2013).

Most *Pinus* spp. are susceptible to DNB (Fraser et al. 2015), although host resistance increases with age (Bulman et al. 2013). In the more recent outbreaks of DNB, however, severe disease has been observed on older trees, suggesting that there are either (i) different genotypes of the pathogen being spread, or due to greater sexual reproduction occurring, or (ii) changing climatic conditions have reduced the abilities of trees to resist infection. Genetically controlled resistance to DNB is known in several *Pinus* spp. (e.g. Ivory and Paterson 1970; Bulman et al. 2013; Fraser et al. 2015) and may be exploited in the future to maintain plantations of *Pinus* spp. in climates known to be conducive to the disease.

Other fungal pathogens related to *Dothistroma* and causing severe needle blights of *Pinus* spp. in tropical plantations include *Cercospora pini-densiflorae* (brown needle disease), attacking *Pinus caribaea*, *P. merkusii* and *P. roxburghii* in Africa, Asia and Central America (Bednářová et al. 2013). This pathogen requires temperatures above 20°C to infect and cause disease, compared with

Figure 7 Symptoms of *Dothistroma septosporum* infection on pines. (a) Young plant with moderate symptoms on older and suppressed needles; (b) typical lesions on mature needles; (c) sparse foliage and needle tip death.

the lower temperatures at which *D. septosporum* and *D. pini* can complete their life cycles. *Lecanosticta acicola*, cause of brown spot needle blight on pines, can be serious in North America, and has spread into Central America, South America (Colombia), and more recently into Europe and South China (Bednářová et al. 2013).

Diplodia shoot blight affects Pinaceae in temperate and tropical regions (Capretti et al. 2013; Fig. 8). Plantation-grown *Eucalyptus* spp. are attacked by a range of *Mycosphaerella* spp., causing leaf-spot diseases, sometimes leading to defoliation (Crous 1998; Crous et al. 2007; Fig. 9). Under certain conditions, *Eucalyptus* in Southern Africa is also prone to a Cylindrocladium shoot blight (Roux and Coetzee 2005; Fig. 10). Often, these diseases cause relatively minor damage to foliage and there are no lasting effects on the hosts: when other pests or disease problems are also present, however, much greater losses may result.

Young foliage of both gymnosperms and angiosperms are also susceptible to rust infections (Kolmer et al. 2009; Hamelin 2013). These pathogens usually occur only when the appropriate alternate hosts are also present in the ecosystem, as the spore type that infects the primary host tree is produced on unrelated alternate hosts. Macrocyclic rusts have a short, repeating life cycle, which rapidly amplifies the amount of inoculum present during the host growing season, as demonstrated for teak leaf rust in Southeast Asia and elsewhere in the tropics caused by *Olivea tectonae* and with the many known species of Acacia rust (Punithalingam and Jones 1971; Mulder and Gibson 1973; Daly et al. 2006).

Figure 8 Severe shoot dieback caused by *Diplodia pini* on *Pinus radiata*.

Figure 9 *Mycosphaerella* leaf spot on *Eucalyptus nitens*.

A particular concern currently in tropical and subtropical regions is myrtle rust, caused by *Austropuccinia psidii* (formerly *Puccinia psidii*; Fig. 11). First recorded attacking guava in Brazil, the pathogen has spread through many regions of the tropics and subtropics where myrtaceous plants are common (Coutinho et al. 1998; Roux et al. 2013; Berthon et al. 2018). The pathogen has a very wide host range. In Australia, Myrtaceae represent approximately 10% of the native flora, and it is estimated that half of these plants grow naturally in regions with climates conducive to myrtle rust infection and development (Berthon et al. 2018).

Shoot and stem-infecting rusts of pines are common in temperate climates (i.e. *Cronartium ribicola*; *Endocronartium harknessii*) but to date have not been recorded in tropical countries. Macrocyclic leaf rusts are also quite common on Pinaceae in natural forests and adjacent planted areas but are

Figure 10 Cylindrocladium shoot blights on (a) seedlings of *Eucalyptus grandis* and (b) *Pinus palustris*. (Images courtesy of Edward L. Barnard, Florida Department of Agriculture and Consumer Services, Bugwood.org).

Figure 11 *Austropuccinia psidii* is causing serious issues with Myrtaceae in tropical and subtropical regions of the world. (a) heavily infected guava fruit; (b) rust lesions on *Melaleuca quinquenervia*; (c) pustules of the pathogen erupting on the foliage of *Eucalyptus*; (d) heavy infection on a young *Eucalyptus* shoot. (Images a,b courtesy of Angus Carnegie; c,d Acelino Alfenas).

usually insignificant. A leaf rust (*Aecidium fragiforme*) is common on *Agathis macrophylla* in Southeast Asia (Punithalingam and Jones 1971).

Powdery mildews are commonly found affecting foliage and green shoots of angiosperms. The main symptom is the presence of a white powdery growth on the foliage or shoot surface, along with growth distortions and, eventually, necrosis. The white powder indicates the production of copious quantities of asexual spores of the pathogen; as the mycelium ages, small brown to black dots (chasmothecia) appear on the host surface, which contain the sexual ascospores. Powdery mildews are common on young foliage of *Eucalyptus* spp., *Acacia* spp. (Fig. 12), *Tectona grandis* and many other trees and may occasionally cause significant disease on mature trees (Limkaisang et al. 2006).

6 Virus diseases

Essentially, viruses cause disease by hijacking the host metabolic machinery, converting cells to virus replication and assembly. Viruses themselves have little

Figure 12 Powdery mildew symptoms on Acacia seedlings. (Courtesy of L. Su See).

or no metabolic machinery, comprising nucleic acids, either RNA or DNA, along with satellite nucleic acids encoding proteins that support replication (Agrios 2005). On infection of a suitable host, virus particles lose the protein coat, and the virus nucleic acid takes over plant cell metabolism leading to replication. Following replication, the protein coat (capsid) is assembled around the replicated DNA or RNA, and the host cell dies. Viroids are another group of infectious agents, existing as naked RNA, lacking a capsid (Agrios 2005).

The literature on virus diseases of trees is limited, although it is likely that these long-lived organisms have many infections (Cooper 1993; Büttner et al. 2013). Problems become obvious when the infected plants deviate from 'normal' appearance, showing symptoms such as chlorosis (mosaics, mottling, yellowing, flecking), growth distortions (galls, witches' brooms, stem pitting), necrosis (either localized or systemic) or wilting of the whole plant. Virus diseases may affect root systems. Hollings (1983) suggested that cacao swollen shoot virus caused swelling and necrosis of *Theobroma cacao* roots.

Viruses cannot penetrate plant cell walls, mostly relying on vectors to damage the host; transmission may also occur via root-grafting, in seed or pollen (Cooper 1993; Hollings 1983). The most common vectors are Arthropoda, with sap-sucking insects being the major culprits; mites, nematodes and fungi may also vector viruses. Human activities including vegetative propagation or pruning are also important in horticulture.

Suspected virus diseases are commonly diagnosed using serologically based enzyme-linked immunosorbant assay (ELISA) tests, or polymerase chain reaction and sequencing (PCR; Büttner et al. 2013). ELISA is based on antibodies against known viruses, which exist for the most serious diseases of agricultural and horticultural crop plants (Cambra et al. 2006), but these tests would not detect unknown viruses. Much more research is required to develop a true understanding of the problems caused by viruses on trees.

Control of virus diseases is difficult, particularly in perennial plants. Quarantine measures attempt to exclude viruses or known vectors from non-invaded environments (Thresh 2003). For some high-value agricultural and horticultural crops, resistant or tolerant cultivars are available (Agrios 2005). Additional approaches include meristem tip culture to eliminate the virus from the host, along with thermotherapy at temperatures where the plant grows, but virus replication is inhibited through antiviral chemotherapy. Propagation and pruning procedures based on routine hygiene are fundamental in horticulture to prevent spread.

Possible viral infections suggested on tropical broadleaved trees including *Acacia* spp., *Cassia* spp., *Paraserianthes falcataria* and *Swietenia macrophylla* (Gibson 1975), but the diagnosis was based on symptoms alone: no stringent diagnoses were carried out. Tree diseases caused by viroids are unknown, although a coconut disease, cadang-cadang, is probably a viroid infection (Vadamalai et al. 2017; Semancik et al. 1987).

7 Nematodes

The most significant nematode problem affecting trees is undoubtedly pine wilt disease, caused by *Bursaphelenchus xylophilus*. The problem is most damaging in warm temperate climates but is also present in subtropical regions of the Far East (Kamata and Takeuchi 2013). The nematode is native in North America, where it is saprotrophic on indigenous Pinaceae. It is vectored between hosts by long-horned beetles in the genus *Monochamus* and has established in various regions, including temperate and subtropical areas in Asia and the west of the Iberian Peninsula.

Another *Bursaphelenchus* species, *B. coccophilus*, probably native in South America, has spread into Central America and the Caribbean, causing mortality on palms including coconut, African oil palm and date palm. Infection results in a red ring observable in cross-sections of affected plants. As *B. coccophilus* is vectored by *Rhyncophorus* spp., it may spread rapidly in southern Europe where the red palm weevil, *R. ferrugineus*, is causing massive mortality of ornamental palms (Kamata and Takeuchi 2013).

8 Algae and parasitic plants

A widespread disease on a wide range of tropical and subtropical woody plants, erroneously known as red rust, is caused by an algal species, *Cephaleuros virescens*. Hosts include tea, cocoa, rubber and forest trees such as *Albizia* spp., *Cassia* spp., *Khaya* spp., *Swietenia macrophylla*, *Tectona grandis* and *Lophostemon confertus* (*Tristania conferta*) (http://www.pestnet.org/fact_sheets/algal_leaf_spot_148.htm). The disease is most common in hot, humid climates, affecting the foliage of heavily shaded, suppressed branches.

Various angiosperm parasitic plants attack trees, the most significant of which are in the Loranthaceae (mistletoes), the Convolvulaceae (Dodders) and the Santalaceae (sandalwoods) (Waller et al. 2001; Shaw and Mathiasen 2013).

Several groups of parasitic plants are hemi-parasites, containing chlorophyll enabling the production of sugars, but gaining water and minerals from a host plant (Parker and Riches 1993). Water and nutrients are taken into the parasite via a specialized haustorium that penetrates the stem or branch and abstracts requirements from the xylem of the host plant.

There are two major groups of mistletoes: the dwarf mistletoes (*Arceuthobium* spp.) and the true or leafy mistletoes in several genera (see later in this chapter). Dwarf mistletoes (family: Santalaceae) parasitize coniferous trees (Hawksworth and Wiens 1996). They are particularly diverse in North America, as far south as the native pine forests of Central America, although there are a few species in Far East Asia and one in Europe (Hawksworth and Wiens 1996; Waller et al. 2001). *Arceuthobium* spp. have minimal true foliage and an explosive dispersal mechanism able to disperse seed over distances as great as 10 m.

True or leafy mistletoes are placed in various families, including the Santalaceae and Loranthaceae. Leafy mistletoes affect many angiosperms and gymnosperms throughout the world, with species of *Amyema, Dendropemon, Dendrophthoe, Elytrante, Loranthus, Macrosolen, Phoradendron, Phthirusa, Psittacanthus, Struthanthus* and *Tapinanthus* common in tropical and subtropical regions (Watson 2001). Large green shoots form conspicuous clusters in tree crowns. The sticky seeds are dispersed by animals, mainly birds. True mistletoes also produce a haustorium into host plant branches, resulting in a swollen, gall-like structure. Significant damage occurs on some affected trees, such as by *Loranthus parasiticus* on *Gmelina arborea* in Bangladesh (Ivory, unpublished) or on *Tectona grandis* (Waller et al. 2001). *Tapinanthus globiferus* (Loranthaceae) has been reported causing severe damage to *Boswellia papyrifera* managed for the production of frankincense in northern Ethiopia (Yirgu et al. 2014).

A number of tree or shrub-sized Santalaceae are partial parasites on a wide range of woody plants, the best-known species being *Santalum album*, which produces a heavily scented timber. Santalaceae species are found from India and Nepal, through Indonesia, Australia and Polynesia. The plants photosynthesize efficiently but are parasitic on the roots of other tree species to increase water and mineral supply (Fig. 13a). Suitability of host species for parasitism by Santalaceae varies; young sandalwood trees are particularly dependent on their hosts, however (Barrett and Fox 1995).

Figure 13 Parasitic plants. (a) A quandong (*Santalum acuminatum*) appears to be an independent shrub/short tree, but is parasitic on the roots of a nearby *Eucalyptus*; (b) typical excess growth of *Cuscuta* spp. on *Cupressus*.

Dodders (*Cuscuta* spp.; Convolvulaceae) are climbing or trailing, leafless, yellow- or orange-coloured plants which spread across aerial parts of herbaceous and woody plants, including broadleaved and gymnosperm trees (Fig. 13b; Parker and Riches 1993). Haustoria are produced into the host stems, removing water and nutrients from the host vascular tissues. The host plant may be weakened by the dodder; a further problem is when dodder growth is abundant, when the host plant may be physically smothered.

9 Disease management

Unlike in agricultural and horticultural crops, disease management methods available to foresters are greatly restricted by factors such as cost, efficacy in the varied environment of the forest ecosystem and non-target effects on beneficial organisms.

Optimal management ideally requires a good understanding of the pathogen life cycle; hence correct diagnosis is crucial. The greatest effect on disease results when management targets a weak point in the pathogen life cycle. The specific infection court, therefore, should be targeted. Researching the life cycle of the disease is time-consuming, but is more cost-effective, both financially and environmentally, than taking a non-targeted approach. With trees in tropical forests, however, knowledge of pathogen life cycles is very limited to date, although the excellent work of scientists at the Forestry and Agriculture Biotechnology Institute in Pretoria, South Africa, is having a great impact on these deficiencies.

Establishment of alien invasive species is increasing on all continents where trees are grown (Santini et al. 2018). Measures applied in an attempt to reduce the likelihood of alien pests and pathogens spreading include legislation and plant quarantine, seed disinfection, prohibition/quarantine of imports and good hygiene in plant nurseries and plantations. Once an invasive organism is established in a hitherto uninvaded (naïve) ecosystem, however, it is usually impossible to eradicate.

Diseases can be avoided by processes such as eradication of an introduced pathogen, or by replanting an affected site with non-host species or provenances of the same species with greater resistance to the pathogen. Certain types of diseases with limited ability to spread over long distances may be avoided by planting sites remote from the outbreak.

If an invasive pathogen is identified early in the establishment phase, it could be eradicated from the newly invaded environment through removal and destruction of any known susceptible hosts within a demarcation zone around the infection centre, followed by regular surveys, to ascertain the success of eradiation. These efforts have rarely been effective, however, except in instances where initial identification was rapidly followed by thorough eradication efforts.

Research using advanced molecular biology techniques has recently shown that tree seed routinely contains numerous fungi, including pathogens (Cleary et al. 2019; Franic et al. 2019). It has long been known that certain *Fusarium* species are associated with tree seed, particularly the potentially highly damaging *Fusarium circinatum*, cause of pine pitch canker (Wingfield et al. 2008b). It is essential, therefore, that seed used in forest nurseries is from certified sources, ideally registered seed stands.

The principle of ecosystem balance between different organisms is the process exploited in biological control. In terms of the biological control of plant diseases, the most common antagonists used have been particular fungi and/or bacteria. Arguably, the most effective biological control agent used in forestry is the application of the decay fungus *Phlebiopsis gigantea* to prevent the colonization of conifer stumps by *Heterobasidion* species, a root and butt rotting Agaricomycete of great importance in temperate forest ecosystems and plantations (Holdenrieder and Greig 1998). In Sumatra, recent work has examined the potential use of another *Phlebiopsis* spp. to prevent *Ganoderma philippii* infecting fresh tree stumps (Agustini et al. 2014). A parallel system was developed in Queensland against the root-killing pathogen *Phellinus noxius* by inoculating *Araucaria cunninghamii* stumps with *Trametes versicolor* and a *Tyromyces* spp. (Anonymous 1988).

Chemical control utilizes three basic approaches: soil sterilants and/or fumigants prior to sowing seed, protectants to reduce the probability of pathogens infecting plants and therapeutants, usually systemic, which are absorbed into the plant and subsequently translocated. Application of xenobiotic agrochemicals to reduce the impacts of pathogens, however, should be used only when other methods of control have failed, and the trees are of sufficient value to warrant the use of expensive treatments.

A further approach to disease control is selection and breeding for resistance. With many agricultural crops, breeding for pest and disease resistance has proven successful, producing hybrids with varying levels of resistance to pathogens, such as rusts, mildews and blights. More recent breeding work has focused on incorporating polygenic resistance into desirable plants, instead of relying on single-gene or oligogenic resistance, which can quickly be overcome by rapidly evolving pathogens. As trees have life cycles of tens to hundreds of years, planting cultivars with polygenic resistance is a far safer option than using monogenic or oligogenic resistance.

Forestry generally utilizes the inherent genetic variability in a tree population, but clonal forestry has some importance in tropical regions, particularly the use of fast-growing hybrids of *Eucalyptus*. Planting mixtures of cultivars, multilines with different disease-resistant genes, is a method to circumvent this problem, although it is recognized that mixtures may lead to problems during harvesting.

10 Where to look for further information

Our knowledge of the disease problems affecting tropical trees is fragmented, with substantial knowledge on a few diseases, but little or only broad knowledge of other pathogens. It is undoubtedly true that many diseases and the causal agent remain to be found in the tropics too. Several institutions are involved in research in these areas, and a few are listed below. We are sure that many other institutions are carrying out good work on diseases of trees in the tropics and exclusion from this list should not be considered detrimental in any way.

Great advances are being made currently at the Forestry and Agricultural Biotechnology Institute (FABI), University of Pretoria, South Africa (www.fabinet.up.ac.za/) where a large group of researchers is focused on pests and diseases affecting both plantation and natural forests. The Forest Research Institute of Malaysia (FRIM) is another institute with excellent facilities for studying tree diseases. A large number of publications have come in recent years from various groups based in Brazil, with a lot of research focused on diseases of plantation Eucalyptus. In Costa Rica, a group led by Marcela Arguda at the Instituto Tecnológico de Costa Rica has a research programme on forest protection (Programa Interinstitucional de Protección Forestal).

There are also institutions outside the tropics with important research programmes addressing pathological problems in tropical forests and plantations. For example, a number of institutions in Australia are active in research of relevance to tropical forest pathology, including groups at Forest Health & Biosecurity, Forest Science, New South Wales, the University of Tasmania and both the University of Queensland and the Queensland Department of Agriculture and Fisheries. In the USA, various groups are involved in tropical pathology research, with a particular emphasis currently on problems caused by Ceratocystis in Hawaii being addressed by people from the USDA Forest Service and the University of Hawaii.

11 Acknowledgements

We are very grateful to a number of people who gave permission for their photographs to be used in illustrating this chapter, including Dr Eric Boa (University of Aberdeen), Prof. Acelino Couto Alfenas (Universidade Federal de Viçosa, Brazil), Dr Geoff Pegg (Queensland Department of Agriculture and Fisheries, Australia), Dr Angus Carnegie (NSW Department of Primary Industries, Australia) and Dr Lee Su See (Forest Research Institute of Malaysia).

12 References

Agrios, G. N. (2005). *Plant Pathology*, 5th edn. Academic Press, New York.

Agustini, L., Wahyuno, D., Indrayadi, H. and Glen, M. (2014). In vitro interactions between *Phlebiopsis sp.* and *Ganoderma philippii*. *Forest Pathology* doi: 10.1111/efp.12145.

Al Adawi, A. O., Barnes, I., Khan, I. A., Al Subhi, A. M., Al Jahwari, A. A., Deadman, M. L., Wingfield, B. D. and Wingfield, M. J. (2013). *Ceratocystis manginecans* associated with a serious wilt disease of two native legume trees in Oman and Pakistan. *Australasian Plant Pathology* 42(2): 179–193.

Anonymous (1988). Biological breakdown of pine stumps. Annual Report 1987–1988. Department of Forestry, Queensland, 27 pp.

Arenz, F. and Simpson, J. A. (1986). Distribution of *Phytophthora cinnamomi* in Papua New Guinea and notes on its origin. *Transactions of the British Mycological Society* 87(2): 289–295.

Arias, S. G., Pons, R. R. and Von Stowasser, E. S. (2013). Temporal analysis of charcoal root rot in forest nurseries under different pathogen inoculum densities and soil moisture content. *Tropical Plant Pathology* 38(3): 179–187.

Ariffin, D., Idris, A. S. and Singh, G. (2000). Status of Ganoderma in oil palm. pp. 49–68. In: Flood, J., Bridge, P. D. and Holderness, M. (Eds). *Ganoderma Diseases of Perennial Crops*. CABI Publishing, Wallingford, New York, London.

Augsperger, C. K. (1983). Seed dispersal of the tropical tree, *Platypodium elegans*, and the escape of its seedlings from fungal pathogens. *Journal of Ecology* 71(3): 759–771.

Barnard, E. L. (1994). The nursery to field carryover and post outplanting impact of *Macrophomina phaseolina* on loblolly pine on a cutover forest site in North Central Florida. *Tree Planter's Notes* 45: 68–71.

Barrett, D. R. and Fox, J. E. D. (1995). Geographical distribution of Santalaceae and botanical characteristics of species in the genus *Santalum*. pp. 3-23. In: Cjerum, L., Fox, J. E. D., Ehrhart, Y. (Eds). *Sandalwood Seed Nursery and Plantation Technology*. Proceedings of the a Regional Workshop for Pacific Island Countries, 1994, New Caledonia. RAS Field Document 8. South Pacific Forestry Development Programme, Suva, Fiji,.

Bednářová, M., Dvořác, M., Janoušek, J. and Jankowský, L. (2013). Other foliar diseases of coniferous trees. pp. 458--487. In: Gonthier, P. and Nicolotti, G. (Eds). *Infectious Forest Diseases*. CABI Publishing, Wallingford, London.

Bergot, M., Cloppet, E., Perarnaud, V., Deque, M., Marcais, B. and Desprez-Loustau, M. L. (2004). Simulation of potential range expansion of oak disease caused by *Phytophthora cinnamomi* under climate change. *Global Change Biology* 10: 1539-1552.

Berthon, K., Esperon-Rodrigues, M., Beaumont, L., Carnegie, A. J. and Leishman, L. (2018). Assessment and prioritisation of plant species at risk from myrtle rust (*Austropuccinia psidii*) under current and future climates in Australia. *Biological Conservation* 218: 247-253.

Boddy, L. and Rayner, A. D. M. (1983). Origins of decay in living deciduous trees: the role of moisture content and a re-appraisal of the expanded concept of tree decay. *New Phytologist* 94(4): 623-641.

Boddy, L. and Watkinson, S. C. (1995). Wood decomposition, higher fungi, and their role in nutrient redistribution. *Canadian Journal of Botany* 73(S1): 1377-1383.

Boerboom, J. H. A. and Maas, P. W. T. (1970). Canker of *Eucalyptus grandis* and *E. saligna* in Surinam caused by *Endothia havanensis*. *Turrialba* 20: 94-99.

Bulman, L. S., Dick, M. A., Ganley, R. J., McDougal, R. L., Schwelm, A. and Bradshaw, R. E. (2013). Dothistroma needle blight. pp. 436--457. In: Gonthier, P. and Nicolotti, G. (Eds). *Infectious Forest Diseases*. CABI Publishing, Wallingford, London.

Büttner, C., van Barge, S., Bandte, M. and Muhlbach, H.-P. (2013). Forest diseases caused by viruses. pp. 50--75. In: Gonthier, P. and Nicolotti, G. (Eds). *Infectious Forest Diseases*. CABI Publishing, Wallingford, London.

Cambra, M., Boscia, D., Myrta, A., Palkocivs, L., Navrátil, M., Barba, M., Gorris, M. T. and Capote, N. (2006). Detection and characterization of plum pox virus: serological methods. *EPPO Bulletin* 36(2): 254-261.

Capretti, P., Santini, A. and Solheim, H. (2013). Branch and tip blights. pp. 420-435. In: Gonthier, P. and Nicolotti, G. (Eds). *Infectious Forest Diseases*. CABI Publishing, Wallingford, London.

Cleary, M., Oskay, F., Doğmuş, H. T., Lehtijärvi, A., Woodward, S. and Vettraino, A. M. (2019). Cryptic risks to forest biosecurity associated with the global movement of commercial seed. *Forests* 10(5): 459 doi: 10.3390/f10050459.

Colquhoun, I. J. and Hardy, G. E. St. J. (2000). Managing the risks of *Phytophthora* root and collar rot during bauxite mining in the *Eucalyptus marginata* (jarrah) forest of Western Australia. *Plant Disease* 84(2): 116-127.

Conradie, E., Swart, W. J. and Wingfield, M. J. (1990). *Cryphonectria* canker of *Eucalyptus*, an important disease in plantation forestry in South Africa. *South African Forestry Journal* 152(1): 43-49.

Cooper, J. I. (1993). *Virus Diseases of Trees and Shrubs*. Chapman & Hall, London.

Coutinho, T. A., Wingfield, M. J., Alfenas, A. C. and Crous, P. W. (1998). Eucalyptus rust: a disease with the potential for serious international implications. *Plant Disease* 82(7): 819-825.

Crous, P. J. (1998). *Mycosphaerella spp. and Their Anamorphs Associated with Leaf Spot Diseases of Eucalyptus*. APS Press, St. Paul, MN.

Crous, P. J. (2002). *Taxonomy and Pathology of Cylindrocladium (Calonectria) and Allied Genera*. APS Press, St. Paul, MN.

Crous, P. J., Summerell, B. A., Carnegie, A. J., Mohammed, C., Himaman, W. and Groenewald, J. Z. (2007). Foliicolous *Mycosphaerella* spp. and their anamorphs on *Corymbia* and *Eucalyptus*. *Fungal Diversity* 26: 143–185.

Daly, A. M., Shivas, R. G., Pegg, G. S. and Mackie, A. E. (2006). First record of teak leaf rust (*Olivea tectonae*) in Australia. *Australasian Plant Disease Notes* 1(1): 25–26.

De Cock, A. W. A. M., Lodhi, A. M., Rintoul, T. L., Bala, K., Robideau, G. P., Abad, Z. G., Coffey, M. D., Shahzad, S. and Levesque, C. A. (2015). *Phytopythium*: molecular phylogeny and systematics. *Persoonia* 34: 25–39.

Drenkhan, R., Tomešová-Haataja, V., Fraser, S., Bradshaw, R. E., Vahalík, P., Mullett, M. S., Martín-García, J., Bulman, L. S., Wingfield, M. J., Kirisits, T., Cech, T. L., Schmitz, S., Baden, R., Tubby, K., Brown, A., Georgieva, M., Woods, A., Ahumada, R., Jankovsky, L., Thomsen, I. M., Adamson, K., Marçais, B., Vuorinen, M., Tsopelas, P., Koltay, A., Halasz, A., La Porta, N., Anselmi, N., Kiesnere, R., Markovskaja, S., Kacergius, A., Papazova-Anakieva, I., Risteski, M., Sotirovski, K., Lazarevic, J., Solheim, H., Boron, P., Bragança, H., Chira, D., Musolin, D. L., Selikhovkin, A. V., Bulgakov, T. S., Keca, N., Karadzic, D., Galovic, V., Pap, P., Markovic, M., Poljakovic Pajnik, L., Vasic, V., Ondrušková, E., Piškur, B., Sadikovic, D., Diez, J. J., Solla, A., Millberg, H., Stenlid, J., Angst, A., Queloz, V., Lehtijärvi, A., Doğmuş-Lehtijärvi, H. T., Oskay, F., Davydenko, K., Meshkova, V., Craig, D., Woodward, S. and Barnes, I. (2016). Global geographic distribution and host range of *Dothistroma* species: a comprehensive review. *Forest Pathology* 46(5): 408–442.

Dublin, H. T. (1995). Vegetation dynamics in the Serengeti-Mara ecosystem: the role of elephants, fire and other factors. pp. 71–90. In: Sinclair, A. R. E. and Arcese, P. (Eds). *Serengeti II: Dynamics, Management, and Conservation of an Ecosystem*. University of Chicago Press, Chicago, IL.

Firmino, A. C., Tozze Jr., H. J., Furtado, E. L. (2012). First report of *Ceratocystis fimbriata* causing wilt in *Tectona grandis* in Brazil. *New Disease Reports* 25: 24.

Franic, I., Prospero, S., Hartmann, M., Allan, E., Auger-Rozenberg, M. A., Grünwald, N. J., Kenis, M., Roques, A., Schneider, S., Sniezko, R., Williams, W. and Eschen, R. (2019). Are traded forest tree seeds a potential source of non-native pests? *Ecological Applications* 29(7): e01971.

Fraser, S., Martin-Garcia, J., Perry, A., Kabir, M. S., Owen, T., Solla, A., Doğmuş, H. T., Brown, A. V., Bulman, L., Barnes, I., Hale, M. D., Vasconcelos, M. W., Lewis, K. J., Woodward, S. and Bradshaw, R. E. (2015). A review of Pinaceae resistance mechanisms against needle and shoot pathogens with a focus on the *Dothistroma-Pinus* interaction. *Forest Pathology* 46: 453–471.

Gibson, I. A. S. (1975). *Diseases of Forest Trees Widely Planted as Exotics in the Tropics and Southern Hemisphere. Part 1. Important Members of the Myrtaceae, Leguminosae, Verbenaceae and Meliaceae*. CMI, Kew/CFI, Oxford, 51 pp.

Hadfield, J. S., Goheen, D. J., Filip, G. M., Schmitt, C. L. and Harvey, R. D. (1986). Root diseases in Oregon and Washington conifers. Report No. R6-FPM-250-86. USDA Forest Service, Pacific Northwest Region, Portland, OR, 27 pp.

Hamelin, R. (2013). Tree rusts. pp. 547-566. In: Gonthier, P. and Nicolotti, G. (Eds). *Infectious Forest Diseases*. CABI Publishing, Wallingford, London.

Hansen, E. M., Reeser, P. W. and Sutton, W. (2012). *Phytophthora* Beyond agriculture. *Annual Review of Phytopathology* 50: 359-378.

Harrington, T. C. (2013). *Ceratocystis* diseases. pp. 230-255. In: Gonthier, P. and Nicolotti, G. (Eds). *Infectious Forest Diseases*. CABI Publishing, Wallingford, London.

Hawksworth, F. G. and Wiens, D. (1996). *Dwarf Mistletoes: Biology, Pathology and Systematics*. Agricultural Handbook 709, Forest Service, USDA, Washington, DC.

Hodges, C. S. (1980) The taxonomy of *Diaporthe cubensis*. *Mycologia* 72: 542-548.

Holdenrieder, O. and Greig, B. J. W. (1998). Biological control. pp. 235-258. In: Woodward, S., Stenlid, J., Karjalainen, R. and Huttermann, A. (Eds). *Heterobasidion Annosum: Biology, Ecology, Impact and Control*. CABI Publishing, Wallingford, London.

Hollings, M. (1983). Virus diseases. pp. 46-77. In: Johnston, A. and Booth, C. (Eds) *Plant Pathologist's Pocket Book*, 2nd edn. Commonwealth Mycological Institute, Kew.

Ivory, M. H. and Paterson, D. N. (1970). Progress in breeding *Pinus radiata* resistant to *Dothistroma* needle blight in East Africa. *Silvae Genetica* 19: 38-42.

Jacobs, K. and Wingfield, M. J. (2001). *Leptographium Species: Tree Pathogens, Insect Associates, and Agents of Blue-Stain*. APS Press, St. Paul, MN, 207 pp.

Jung, T., Orlikowski, L., Henricot, B., Abad-Campos, P., Aday Kaya, A. G., Aguín Casal, O., Bakonyi, J., Cacciola, S. O., Cech, T., Chavarriaga, D., Corcobado, T., Cravador, A., Decourcelle, T., Denton, G., Diamandis, S., Doğmuş Lehtijärvi, H. T., Franceschini, A., Ginetti, B., Glavendekic, M., Hantula, J., Hartmann, G., Herrero, M., Ivic, D., Horta Jung, M., Lilja, A., Keca, N., Kramarets, V., Lyubenova, A., Machado, H., Magnano di San Lio, G., Mansilla Vázquez, P. J., Marçais, B., Matsiakh, I., Milenkovic, I., Moricca, S., Nagy, Z. Á., Nechwatal, J., Olsson, C., Oszako, T., Pane, A., Paplomatas, E., Pintos Varela, J. C., Prospero, S., Rial Martínez, C., Rigling, D., Robin, C., Rytkönen, A., Sánchez, M. E., Scanu, B., Schlenzig, A., Schumacher, J., Slavov, S., Solla, A., Sousa, E., Stenlid, J., Talgø, V., Tomic, Z., Tsopelas, P., Vannini, A., Vettraino, A. M., Wenneker, M., Woodward, S. and Peréz-Sierra, A. (2015). Widespread *Phytophthora* infestations in European nurseries put forest,semi-natural and horticultural ecosystems at high risk of Phytophthora diseases. *Forest Pathology* 46: 134-163.

Kamata, N. and Takeuchi, Y. (2013). Pine wilt and other nematode diseases. pp. 115-127. In: Gonthier, P. and Nicolotti, G. (Eds). *Infectious Forest Diseases*. CABI Publishing, Wallingford, London.

Keane, P. J., Kyle, K. A., Podger, F. D. and Brown, B. N. (2001). *Diseases and Pathogens of Eucalypts*. CSIRO Publishing, Victoria, Australia.

Kolmer, J. A., Ordonez, M. and Groth, J. V. (2009). The rust fungi. In: *Encylopedia of Life Sciences (ELS)*. John Willey & Sons, Ltd, Chichester. doi: 10.1002/9780470015902.a0021264.

Lee, S. S. (2004). Diseases and potential threats to *Acacia mangium* plantations in Malaysia. *Unasylva* 217: 31-35.

Limkaisang, S., Cunnington, J. H., Liew, K. W., Salleh, B., Sato, Y., Divarangkoon, R., Fangfuk, W., To-anun, C. and Takamatsu, S. (2006). Molecular phylogenetic analyses reveal close relationship of powdery mildew fungi on some tropical trees with *Erysiphe alphitoides*, an oak powdery mildew. *Mycoscience* 47: 327-335.

Lundquist, J. E. (1987). A history of five forest diseases in South Africa. *South African Forestry Journal* 140(1): 51-59.

Mohammed, C. L., Rimbawanto, A. and Page, D. E. (2014). Management of basidiomycete root- and stem-rot diseases in oil palm, rubber and tropical hardwood plantation crops. *Forest Pathology* 44(6): 428-446.

Mulder, J. L. and Gibson, I. A. S. (1973). Olivea tectonae. *CMI Descriptions of Pathogenic Fungi and Bacteria*, No. 365. Commonwealth Mycological Institute, Kew, UK.

Old, K. M. and Santos Cristovao, C. D. (2003). A rust epidemic of the coffee shade tree (*Paraserianthes falcataria*) in East Timor. pp. 139–145. In: da Costa, H., Piggin, C., da Cruz, C. J. and Fox, J. J. (Eds). *Agriculture: New Directions for a New Nation - East Timor (Timor-Leste)*. ACIAR Proceedings, No. 113. Canberra, Australia.

Packer, A. and Clay, K. (2000). Soil pathogens and spatial patterns of seedling mortality in a temperate tree. *Nature* 404(6775): 278–281.

Pardo Cardona, V. M. (1998). Uredinales (Royas) de Cordia L. (Boraginaceae) en Colombia. *Revista Facultad Nacional de Agronomía Medellín* 51: 277–283.

Parfitt, D., Hunt, J., Dockrell, D., Rogers, H. J. and Boddy, L. (2010). Do all trees carry the seeds of their own destruction? PCR reveals numerous wood decay fungi latently present in sapwood of a wide range of angiosperm trees. *Fungal Ecology* 3(4): 338–346.

Parker, C. and Riches, C. R. (1993). *Parasitic Weeds of the World: Biology and Control*. CABI Publishing, Wallingford, Oxford.

Pearce, R. B. (1996). Antimicrobial defences in the wood of living trees. *New Phytologist* 132(2): 203–233.

Punithalingam, E. and Jones, D. (1971). *Aecidium* species on *Agathis*. *Transactions of the British Mycological Society* 57(2): 325–333.

Rajput, N. A., Pathan, M. A., Jiskani, M. M., Rajput, A. Q. and Arain, R. R. (2011). Pathogenecity and host range of *Fusarium solani* (MART) SACC. Causing dieback of shisham (Dalbergia sissoo ROXB.). *Pakistan Journal of Botany* 40: 2631-2639.

Rayner, A. D. M. and Boddy, L. (1988). *Fungal Decomposition of Wood: Its Biology and Ecology*. John Wiley Press, Chichester, New York, Brisbane, Toronto, Singapore.

Roux, J. and Coetzee, M. P. A. (2005). First report of pink disease on native trees in South Africa and phylogenetic placement of *Erythricium salmonicolor* in the Homobasidiomycetes. *Plant Disease* 89(11): 1158–1163.

Roux, J., Greyling, I., Coutinho, T. A., Verleur, M. and Wingfield, M. J. (2013). The myrtle rust pathogen, *Puccinia psidii*, discovered in Africa. *IMA Fungus* 4(1): 155–159.

Santini, A., Ghelardini, L., De Pace, C., Desprez-Loustau, M. L., Capretti, P., Chandelier, A., Cech, T., Chira, D., Diamandis, S., Gaitniekis, T., Hantula, J., Holdenrieder, O., Jankovsky, L., Jung, T., Jurc, D., Kirisits, T., Kunca, A., Lygis, V., Malecka, M., Marçais, B., Schmitz, S., Schumacher, J., Solheim, H., Solla, A., Szabò, I., Tsopelas, P., Vannini, A., Vettraino, A. M., Webber, J., Woodward, S. and Stenlid, J. (2013). Biogeographic patterns and determinants of invasion by alien forest pathogenic fungi in Europe. *New Phytologist* 197(1): 238–250.

Santini, A., Liebhold, A., Migliorini, D. and Woodward, S. (2018). Tracing the role of human civilization in the globalization of plant pathogens. *The ISME Journal* 12(3): 647–652.

Semancik, J. S., Garnsey, S. M., Robertson, H. D. and Symons, R. H. (1987). *Viroids and Viroid-Like Pathogens*. CRC Press, Boca Raton, FL, 192 pp.

Shaw, D. C. and Mathiasen, R. L. (2013). Forest disease caused by higher plants: mistletoes. pp. 97–114. In: Gonthier, P. and Nicolotti, G. *Infectious Forest Diseases*. CABI Publishing, Wallingford, London.

Shearer, B. L., Crane, C. E. and Cochrane, A. (2004). Quantification of the susceptibility of the native flora of the South-West Botanical Province, Western Australia, to *Phytophthora cinnamomi*. *Australian Journal of Botany* 52(4): 435–443.

Shigo, A. L. and Marx, H. G. (1977). Compartmentalization of decay in trees. *USDA Forest Service Agriculture Information Bulletin* 405: 73 pp.

Thresh, J. M. (2003). Control of plant virus diseases in sub-Saharan Africa: the possibility and feasibility of an integrated approach. *African Crop Science Journal* 11(3): 199-223.

Tsui, K. M., Fryar, S. C., Hodgkiss, U., Hyde, K. D., Poonyth, A. D. and Taylor, I. E. (1998). The effect of human disturbance on fungal diversity in the tropics. *Fungal Diversity* 1: 19-26.

Vadamalai, G., Thanarajoo, S. S., Joseph, H., Kong, L. L. and Randles, J. W. (2017). Coconut cadang-cadang viroid and coconut tinangaja viroid. pp. 263-273 in Hadidi, A., Flores, R., Randles, J. W. and Palukaitis, P. (Eds). *Viroids and Satellites*. Academic Press, Cambridge, Mass..

Van Noorwijk, M., Rahayu, S., Williams, S. E., Hairiah, K., Khasan, N. and Schroth, G. (2004). Crop and tree root-system dynamics. Chapter 5, 429 pp. In: van Noorewijk, M., Cadish, G. and Ong, C. K. (Eds). *Below-Ground Interactions in Tropical Agroecosystems, Concepts and Models with Multiple Plant Components*. CABI Publishing, Wallingford, London, New York, ISBN-0851996736.

Waller, J. M., Lenne, J. M. and Waller, S. J. (2001). *Plant Pathologist's Handbook*. CABI Publsishing, Wallingford, London, 528 pp.

Watson, D. M. (2001). Mistletoe - a keystone resource in forests and woodlands worldwide. *Annual Review of Ecology and Systematics* 32(1): 219-249.

Weste, G. (2003). The dieback cycle in Victorian Forests: a 30-year study of changes caused by *Phytophthora cinnamomi* in Victorian open forests, woodlands and heathlands. *Australasian Plant Pathology* 32(2): 247-256.

Wingfield, M. J., Slippers, B., Hurley, B. P., Coutinho, T. A., Wingfield, B. D. and Roux, J. (2008a). Eucalypt pests and diseases: growing threats to plantation productivity.*Southern Forests* 70(2): 139-144.

Wingfield, M. J., Hammerbacher, A., Ganley, R. J., Steenkamp, E. T., Gordon, T. R., Wingfield, B. D. and Coutinho, T. A. (2008b). Pitch canker caused by *Fusarium circinatum* - a growing threat to pine plantations and forests worldwide. *Australasian Plant Pathology* 37(4): 319-334.

Yirgu, A., Gezahgne, A., Kassa, H. and Tsega, M. (2014). Parasitic plant in natural *Boswellia papyrifera* stands at Humera, Northern Ethiopia. *Journal of Forestry Research* 25(4): 923-928.

Chapter 3

Advances in understanding and managing insect pests of forest trees

Barbara Bentz, USDA Forest Service, Rocky Mountain Research Station, USA; Pierluigi Bonello, The Ohio State University, USA; Horst Delb, Forest Research Institute of Baden-Württemberg, Germany; Christopher Fettig, USDA Forest Service, Pacific Southwest Research Station, USA; Therese Poland, USDA Forest Service, Northern Research Station, USA; Deepa Pureswaran, Canadian Forest Service, Laurentian Forestry Centre, Canada; and Steven Seybold, USDA Forest Service, Pacific Southwest Research Station, USA

1 Introduction

Forest ecosystems cover roughly one-third of the world's land base, providing vital contributions to people and the planet we inhabit (FAO, 2018). Boreal and temperate forests, which make up almost half of all forested land and include ~1.3 trillion trees (Crowther et al., 2015), are a major source of timber, biomass, and other ecosystem services including clean air and water, hunting, recreation, cultural identities, and global biodiversity and sustainability (Gauthier et al., 2015; Brecka et al., 2018). Through carbon sequestration and storage, forests are also increasingly recognized for mitigation of atmospheric concentrations of carbon dioxide (CO_2) (Dugan et al., 2018) that are contributing to a changing climate (IPCC, 2014). As carbon sinks,

http://dx.doi.org/10.19103/AS.2019.0057.19
Published by Burleigh Dodds Science Publishing Limited, 2020.

forests worldwide absorb the equivalent of ~2 billion tonnes of CO_2 each year (Köhl et al., 2015; FAO, 2018), and ~42% of that carbon is held in living tree biomass (Pan et al., 2011). The future of forests globally, however, is being threatened by a variety of factors including an increased rate of native disturbances due to a changing climate (Fettig et al., 2013a; Allen et al., 2015), and an increased frequency of invasive species introductions as global trade accelerates (Aukema et al., 2010; Brockerhoff and Liebhold, 2017). Disturbances that result in large amounts of tree mortality can reverse the role of forests from carbon sinks to carbon sources, at least in the short term until regrowth occurs (Hansen et al., 2015; Arora et al., 2016). In a rapidly changing climate, consequences of expansive tree mortality could also include altered successional trajectories with long-term type conversions and negative impacts to biodiversity as ecosystem tipping points are crossed in no-analog climates (Fox, 2007; Anderegg et al., 2013; Vose et al., 2018).

In boreal and temperate forests, disturbances caused by native and non-native invasive insects are among the most important mediators of tree mortality (Berner et al., 2017; Liebhold et al., 2017; Mezei et al., 2017). Although native herbivorous insects are considered integral components of forest ecosystems (Mattson and Addy, 1975) and the majority cause no economic damage (Raffa et al., 2015), species within a few genera are considered pests when outbreaks interfere with land management objectives and cause significant ecological and economic impacts (Cooke et al., 2007; Raffa et al., 2009). Globally, between 2003 and 2012, >70 million ha of boreal and temperate forests were affected by insect pests, with more than half occurring in North and Central America (van Lierop et al., 2015).

Many traits that influence forest insect population success are temperature dependent and warming temperatures are causing range expansions and altered outbreak frequencies (Weed et al., 2013; Pureswaran et al., 2018). Insects not considered pests in their native habitats can become pests when introduced into new habitats as they respond to enemy-free (Keane and Crawley, 2002) and defense-free environments (Showalter et al., 2018). Climatic changes are expected to result in continued alterations in the frequency and distribution of native and invasive insect outbreaks (Logan et al., 2003; Forrest, 2016), in addition to stress-related changes in forest composition and structure that influence suitability to insect predators (Kolb et al., 2016; Marini et al., 2017; Lantschner et al., 2019). The interacting effects of climate change on insects and host trees coupled with increasing arrival rates of non-native insects globally (McCullough et al., 2006; Brockerhoff and Liebhold, 2017) are making management of boreal and temperate forests for timber and other ecosystem services increasingly uncertain and challenging. We review current information and key issues for predicting distributions in a changing climate, and for managing native (Tables 1 and 2)

Table 1 Bark beetles (Coleoptera: Curculionidae, Scolytinae) noted in their native ranges for causing conifer tree mortality in Northern Hemisphere boreal and temperate forests

Common name	Scientific name	Native continent	Common host(s)
Arizona fivespined ips	*Ips lecontei*	North America	*Pinus ponderosa*
California fivespined ips	*Ips paraconfusus*	North America	*P. attenuata, P. coulteri, P. lambertiana, P. ponderosa, P. radiata*
Chinese white pine beetle	*Dendroctonus armandi*	Asia	*P. armandii*
Douglas-fir beetle	*Dendroctonus pseudotsugae*	North America	*Pseudotsuga menziesii*, occasionally *Larix occidentalis*
Eastern fivespined ips	*Ips grandicollis*	North America	*P. echinata, P. elliottii, P. taeda, P. virginiana*
Eastern larch beetle	*Dendroctonus simplex*	North America	*Larix laricina*
Fir engraver	*Scolytus ventralis*	North America	*Abies concolor, A. grandis, A. magnifica*
Great spruce bark beetle	*Dendroctonus micans*	Europe, Asia	*Pi. abies; P. sylvestris*
Jeffrey pine beetle	*Dendroctonus jeffreyi*	North America	*P. jeffreyi*
Larger eight-toothed European spruce bark beetle (= European spruce beetle)	*Ips typographus*	Europe, Asia	*Picea abies, Pi. orientalis, Pi. yezoensis*, occasionally *P. sylvestris*
Larger Mexican pine beetle	*Dendroctonus approximatus*	North America	*P. engelmannii, P. leiophylla, P. ponderosa*
Mountain pine beetle	*Dendroctonus ponderosae*	North America	*P. albicaulis, P. contorta, P. flexilis, P. lambertiana, P. monticola, P. ponderosa*
Northern spruce engraver	*Ips perturbatus*	North America	*Pi. engelmannii, Pi. glauca, Pi. x lutzii, occasionally Pi. mariana*
Pine engraver	*Ips pini*	North America	*P. banksiana, P. contorta, P. jeffreyi, P. lambertiana, P. ponderosa, P. resinosa, P. strobus*
Pinyon ips	*Ips confusus*	North America	*P. edulis, P. monophylla*
Roundheaded pine beetle	*Dendroctonus adjunctus*	North America	*P. arizonica, P. engelmannii, P. flexilis, P. leiophylla, P. ponderosa, P. strobiformis*
Sixspined ips	*Ips calligraphus*	North America	*P. echinata, P. elliotti, P. ponderosa, P. taeda, P. virginiana*
Six-toothed bark beetle	*Ips sexdentatus*	Europe, Asia	*P. heldreichii, P. nigra, P. pinaster, P. sylvestris, Pi. orientalis*

(*Continued*)

Table 1 (*Continued*)

Common name	Scientific name	Native continent	Common host(s)
Southern pine beetle	*Dendroctonus frontalis*	North America	*P. echinata, P. engelmannii, P. leiophylla, P. ponderosa, P. rigida, P. taeda, P. virginiana*
Spruce beetle	*Dendroctonus rufipennis*	North America	*Pi. engelmannii, Pi. glauca, Pi. pungens, Pi. sitchensis*
Western balsam bark beetle	*Dryocoetes confusus*	North America	*A. lasiocarpa*
Western pine beetle	*Dendroctonus brevicomis*	North America	*P. coulteri, P. ponderosa*

and invasive (Table 3) forest insects that are considered pests in boreal and temperate forests of the northern hemisphere. Five case studies highlight recent strategies being applied to mitigate forest insect-caused tree mortality (Fig. 1; Section 7).

2 Advances in understanding and predicting native and invasive forest insect responses to climate change

The timing of seasonal life history events (i.e. phenology) is critical for insects, especially in boreal and temperate regions of the world. Strategies are required to survive harsh winters, and the relatively short growing season demands that organisms prepare to feed and reproduce as soon as temperatures become sufficiently warm (Bale et al., 2002). Forest insects have adapted multiple thermally dependent phenological strategies to survive and persist in harsh climates including development rates and thermal thresholds, diapause (a dormant physiological state entered to survive harsh conditions and increase cohort synchrony), and cold-hardening (increased cold tolerance through acclimation and metabolic processes) (Bentz and Jönsson, 2015). These traits, in addition to interactions with host trees and community and symbiotic associates, will determine species- and location-specific responses of forest insects to a changing climate (Netherer and Schopf, 2010; Weed et al., 2013). In northern latitudes, phenological traits of insects have influenced range limits where, despite the availability of host plants, climates were historically too cool to allow completion of life cycles or permit overwinter survival (Safranyik and Wilson, 2006). Northern latitudes are warming more rapidly than global means (IPCC, 2014) and observations of movement to new, thermally suitable environments is the most noticeable response of organisms to climate warming (Chen et al., 2011).

Published by Burleigh Dodds Science Publishing Limited, 2020.

Table 2 Defoliator species that can cause extensive tree mortality in their native habitats in Northern Hemisphere boreal and temperate forests

Common name	Scientific name	Native continent	Common host(s)
Asian gypsy moth	*Lymantria dispar asiatica* Lepidoptera: Erebidae	Asia	*Acer* spp., *Alnus* spp., *Betula* spp., *Larix* spp., *Picea* spp., *Pinus* spp., *Populus* spp., *Quercus* spp.
Autumnal moth	*Epirrita autumnata* Lepidoptera: Geometridae	Europe	*Betula pubescens*
Balsam fir sawfly	*Neodiprion abietis* Hymenoptera: Diprionidae	North America	*Abies balsamea, Picea glauca, Pi. mariana*
European gypsy moth	*Lymantria dispar dispar* Lepidoptera: Erebidae	Europe	*Acer* spp., *Alnus* spp., *Betula* spp., *Populus* spp., *Quercus* spp.
European pine sawfly	*Neodiprion sertifer* Hymenoptera: Diprionidae	Europe	*Pinus* spp.
Forest tent caterpillar	*Malacosoma disstria* Lepidoptera: Lasiocampidae	North America	*Acer saccharum, Populus* spp.
Hemlock looper	*Lambdina fiscellaria* Lepidoptera: Geometridae	North America	*Abies balsamea, Acer saccharum, Betula papyrifera, Picea glauca, Tsuga canadensis*
Jack pine budworm	*Choristoneura pinus* Lepidoptera: Tortricidae	North America	*Pinus banksiana*
Larch bud moth	*Zeiraphera diniana* Lepidoptera: Tortricidae	Europe	*Larix decidua, Pinus cembra*
Nun moth	*Lymantria monacha* Lepidoptera: Erebidae	Europe	*Abies* spp., *Larix* spp., *Picea* spp., *Pinus* spp.
Pine moth	*Dendrolimus spectabilis* Lepidoptera: Lasiocampidae	Asia–Japan	*Pinus densiflora, P. thunbergii*
Pine processionary moth	*Thaumetopoea pityocampa Lepidoptera: Thaumetopoeidae*	Europe	*Pinus* spp.
Redheaded pine sawfly	*Neodiprion lecontei* Hymenoptera: Diprionidae	North America	*Pinus* spp.
Scarce umber	*Agriopis aurantiaria* Lepidoptera: Geometridae	Europe	*Betula* spp., *Quercus* spp., *Salix* spp., *Ulmus* spp.
Siberian silk moth/white-lined silk moth	*Dendrolimus sibiricus/superans* Lepidoptera: Lasiocampidae	Northern Asia	*Abies sibirica, Pinus sibirica, Picea spp., and Larix* spp.
Spruce budworm	*Choristoneura fumiferana* Lepidoptera: Tortricidae	North America	*Abies balsamea, Picea glauca, Pi. mariana*
Western spruce budworm	*Choristoneura freemani* Lepidoptera: Tortricidae	North America	*Picea engelmanni, Pseudotsuga menziesii*
Winter moth	*Operophtera brumata* Lepidoptera: Geometridae	Europe	*Betula pubescens*
Yellowheaded spruce sawfly	*Pikonema alaskensis* Hymenoptera: Tenthredinidae	North America	*Picea* spp.

Table 3 Examples of invasive forest insect pests that cause significant damage in Northern Hemisphere boreal and temperate forests

Common name	Scientific name	Native continent	Introduced range	Hosts	Selected References
Asian gypsy moth	*Lymantria dispar asiatica* Vnukovskij; *Lymantria dispar japonica* (Motchulsky) (Lepidoptera: Erebidae)	Asia	Russia	*Quercus, Betula, Alnus, Salix, Prunus, Larix, Pseudotsuga*	Pogue and Schaefer (2007)
Asian longhorned beetle	*Anoplophora glabripennis* Motchulsky (Coleoptera: Cerambycidae)	Asia	North America, Europe	*Acer, Populus, Salix,* others	Hu et al. (2009), Dodds and Orwig (2011)
Balsam woolly adelgid	*Adelges piceae* Ratzeburg (Hemiptera: Adelgidae)	Europe	North America	*Abies*	Smith and Nicholas (1998)
Beech scale + beech bark disease	*Cryptococcus fagisuga* Lindinger (Hemiptera: Eriococcidae)+ *Nectria coccinea* var. *faginata*, *Nectria galligena*	Europe	North America	*Fagus*	Houston (1994)
Blue gum chalcid	*Leptocybe invasa* Fisher & La Salle (Hymenoptera: Chalcidae)	Australia	Africa, Asia, Europe	*Eucalyptus, Corymbia*	Mendel et al. (2004), Zheng et al. (2014)
Chestnut gall wasp	*Dryocosmus kuriphilus* Yasumatsu (Hymenoptera: Cynipidae)	Asia	Europe and North America	*Castanea*	Rieske (2007), Battisti et al. (2014)
Citrus longhorned beetle	*Anoplophora chinensis* Forster (Coleoptera: Cerambycidae)	China, Korea, Japan	Europe	*Acer, Citrus, Populus, Ulmus,* others	Hérard and Maspero (2019)
Coconut hispine beetle	*Brontispa longissimi* (Gestro) (Coleoptera: Chrysomelidae)	Indonesia	China	*Areca, Cocos, Elaeis, Metroxylon, Phoenix*	Nakamura et al. (2006)
Cypress aphid	*Cinara cupressi sensu lato* (Hemiptera: Aphididae)	North America	Africa, Europe	*Cupressus, Juniperus, Thuja*	Watson et al. (1999), Mendel et al. (2016)
Cypress jewel beetle	*Lamprodila* (*Palmar*) *festiva* (L.) (Coleoptera: Buprestidae)		Russia	*Chamaecyparis, Cupressus, Juniperus, Thuja*	Volkovitsh and Karpun (2017)
Emerald ash borer	*Agrilus planipennis* Fairmaire (Coleoptera: Buprestidae)	Asia	North America, Europe, Russia	*Fraxinus, Olea*	Herms and McCullough (2014)
Eucalyptus leaf weevils	*Gonipterus* spp. (Coleoptera: Curculionidae)	Australia	Africa	*Eucalyptus*	Hurley et al. (2016)

Eucalyptus longhorned beetle	*Phoracantha semipunctata* (F.) (Coleoptera: Cerambycidae)	Australia	North America, Israel, Turkey, Europe	*Eucalyptus*	Scriven et al. (1986), Paine (2016)
European gypsy moth	*Lymantria dispar dispar* (L.) (Lepidoptera: Erebidae)	Europe	North America	*Quercus*, others	Davidson et al. (2001), Tobin and Liebhold (2011)
Fall webworm	*Hyphantria cunea* Drury (Lepidoptera: Erebidae)	North America	China, Europe	*Betula, Carya, Platanus, Salix, Ulmus,* others	Szalay-Marzso (1971), Yang et al. (2008)
Four-eyed fir bark beetle	*Polygraphus proximus* Blandford (Coleoptera: Curculionidae: Scolytinae)	China, Korea, Russian Far East	European Russia	*Abies, Larix, Picea*	Baranchikov et al. (2010)
Hemlock woolly adelgid	*Adelges tsugae* (Annand) (Hemiptera: Adelgidae)	Japan	North America	*Tsuga*	Vose et al. (2013)
Loblolly pine mealybug	*Oracella acuta* (Lobdell) (Hemiptera: Pseudococcidae)	USA	China	*Pinus*	Chen et al. (2017), Lu and Sun (2017)
Pine needle hemiberlesian scale	*Hemiberlesia pitysophila* Takagi (Hemiptera: Diaspididae)	Japan	China	*Pinus*	Feng et al. (2009)
Pine wood nematode + pine wilt disease	*Monochamus* spp. (Coleoptera: Cerambycidae) + *Bursaphelenchus xylophilus*	North America	Asia and Europe	*Pinus*	Lee et al. (2017)
Polyphagous shot hole borer + fusarium dieback	*Euwallacea whitfordiodendrus* and *E. kuroshio* (Coleoptera: Curculionidae: Scolytinae) + *Fusarium* spp., *Acremonium* spp., *Graphium* spp.	Asia	North America, Israel	*Acer, Alnus, Platanus, Populus, Salix,* others	Eskalen et al. (2013), Lynch et al. (2016), Gomez et al. (2018), Coleman et al. (2019)
Red turpentine beetle	*Dendroctomus valens* (LeConte) (Coleoptera: Curculionidae: Scolytinae)	North and Central America	China	*Pinus, Larix, Picea*	Sun et al. (2013)
Redbay ambrosia beetle + laurel wilt	*Xyleborus glabratus* Eichhoff (Coleoptera: Curculionidae: Scolytinae) + *Raffealea lauricola*	India, Japan, Myanmar, and Taiwan	North America	*Persea, Sassafras*, others	Fraedrich et al. (2008), Hughes et al. (2015)
Siberian silk moth	*Dendrolimus sibiricus* Tschetv. (Lepidoptera: Lasiocampidae)	Siberia, China, North Korea	Russia, Central Europe	*Pinus, Larix, Abies, Picea*	Kirichenko et al. (2009), Kononov et al. (2016)

(*Continued*)

Table 3 (*Continued*)

Common name	Scientific name	Native continent	Introduced range	Hosts	Selected References
Sirex woodwasp	*Sirex noctilio* (F.) (Hymenoptera: Siricidae)	Eurasia, North Africa	Australia, New Zealand, North and South America, South Africa	*Pinus*	Haavik et al. (2015)
Smaller European elm and banded elm bark beetles + Dutch elm disease	*Scolytus multistriatus* (Marsham) and *S. schevyrewi* (Semenov) (Coleoptera: Curculionidae: Scolytinae) + *Ophiostoma ulmi, Ophiostoma novo-ulmi*	Europe	North America, New Zealand	*Ulmus*	Brasier and Buck (2001), Negrón et al. (2005), Jacobi et al. (2013)
Spotted lantern fly	*Lycorma delicatula* (White) (Hemiptera: Fulgoridae)	Northern China	North America	*Acer, Ailanthus, Malus, Populus, Prunus,* others	Dara et al. (2015)
Thuja shoot borer	*Phloeosinus aubei* (Perris)	Mediterranean region	Central Europe, expanded range	*Thuja*	Bozsik and Szőcs (2017)
Walnut twig beetle + thousand cankers disease	*Pityophthorus juglandis* Blackman (Coleoptera: Curculionidae: Scolytinae) + *Geosmithia morbida*	North America	North America-expanded range, Europe-Italy	*Juglans, Pterocarya*	Tisserat et al. (2009), Seybold et al. (2016, 2019)
Western conifer seedbug	*Leptoglossus occidentalis* Heidemann (Hemiptera: Coreidae)	North America	Europe, Japan, North America-expanded range	*Abies, Pinus, Pseudotsuga, Tsuga*	McPherson et al. (1990), Ishikawa and Kikuhara (2009), Lesieur et al. (2019)
Winter moth	*Operophtera brumata* L. (Lepidoptera: Geometridae)	Europe	North America	*Acer, Betula, Malus, Ostrya, Quercus, Tilia, Ulmus*	Simmons et al. (2014)
Yellow phoracantha	*Phoracantha recurva* Newman (Coleoptera: Cerambycidae)	Australia, Papua New Guinea	North America	*Eucalyptus*	Hanks et al. (1997), Paine (2016)

Figure 1 Adult and larval form, and forest ecosystem impact of insects described in case studies (Section 7). (a) emerald ash borer, (b) mountain pine beetle, (c) spruce budworm, (d) European spruce beetle, and e) European gypsy moth. See Tables 1-3 for additional information on each species. Photo credits: B. Bentz, M. Ayres; Bugwood.org - D. Adam, D. Cappaert, G. Csoka, W. Ciesla, E. Day, G. Ghent, T. Kimoto, F. Lakatos, M. Zubrik.

2.1 Observed and predicted responses to changing climate

As might be expected, warming winter temperature is most often cited as the key environmental factor influencing observed native forest insect population range shifts or expansions northward (Table 4). Warming spring temperatures that positively influence defoliator larval feeding to be synchronous with bud break of host trees can also be a contributing factor to range shifts (Table 4). While it is clear that warming has allowed some native forest insect populations to expand northward, many populations have also exhibited increased duration and intensity of population outbreaks within historic geographic ranges as a result of the combined effects of drought, increased storms that result in stressed host trees, and warming summer and winter temperatures that reduce generation time and increase population survival (e.g., Berg et al., 2006; Jactel et al., 2012; Flower et al., 2014; Weed et al., 2015; Kolb et al., 2016; Berner et al., 2017; Hart et al., 2017; Marini et al., 2017; Fettig et al., 2019; Ward et al., 2019). Although range expansions or shifts are occurring and warming temperatures have generally been favorable for forest insects, climate change may also restrict native and invasive population success (Forrest, 2016). Indeed, recent observations show that long-term native and invasive population spatial synchrony and outbreak cycles have been disrupted, resulting in negative impacts to forest insect populations (Esper et al., 2006; Johnson et al., 2010; Haynes et al., 2014; Cooke and Roland, 2018).

Table 4 Observed recent range expansions of forest insect species in their native continents associated with abiotic factors. See Tables 1 and 2 for additional species information.

Common name	Native continent	Inciting abiotic factor(s)	References
Autumnal moth	Europe	Warming mean and winter temperature	Jepsen et al. (2008)
Mountain pine beetle	North America	Warming winter temperature	Carroll et al. (2004), Stahl et al. (2006), Sambaraju et al. (2012)
Pine processionary moth	Europe	Warming winter temperature	Battisti et al. (2005), Netherer and Schopf (2010)
Scarce umber	Europe	Warming spring temperature	Jepsen et al. (2011)
Siberian silk moth	Northern Asia	Sum of temperatures >0°C, drought	Kharuk et al. (2018)
Southern pine beetle	North America	Warming winter temperature	Weed et al. (2013), Dodds et al. (2018)
Spruce budworm	North America	Warming spring and winter temperature	Candau and Fleming (2005)
Western spruce budworm	North America	Warming year-round temperatures	Thomson and Benton (2007), Maclauchlan et al. (2018), Régnière and Nealis (2019)
Winter moth	Europe	Warming mean and winter temperature	Jepsen et al. (2008)

As climate change is expected to continue (IPCC, 2014), numerous studies have used transplant experiments and a variety of statistical and process-based models, in association with forecasted temperatures based on Global Climate Models, to predict population success and range expansions beyond historical distributions. With a few exceptions, range expansion northward and increased population success were predicted for most species (Table 5). European gypsy moth (*Lymantria dispar dispar*) expansion was predicted to be limited by winter temperatures in North America where it is invasive, and also in Europe where it is native. Process-based models predicted northward expansions for spruce budworm (*Choristoneura fumiferana*), western spruce budworm (*C. freemani*), and mountain pine beetle (*Dendroctonus ponderosae*), but also southern range retractions (Table 5). In contrast with statistical or correlative models, process-based models include species-specific details on thermally dependent traits that drive seasonality and population success, and in all three species excessive warming at southern range limits was predicted to disrupt diapause and developmental traits that have evolved to synchronize individuals with each other, their hosts, and seasonal environments (Régnière et al., 2012; Bentz et al., 2019; Régnière and Nealis, 2019).

Correlative or statistical approaches between insect occurrence and climate variables, which have been most often used to project future

Table 5 Predictions of range expansion and population success for native and invasive forest insect species in a changing climate. See Tables 1-3 for additional species information

Common name	Prediction(s)	References
Emerald ash borer	Range expansion northward in both North America and Europe where it is invasive	Sobek-Swant et al. (2012b), Valenta et al. (2017), Cuddington et al. (2018)
European gypsy moth	Limited by winter cold in North America where it is invasive, except when insulated by snow cover; limited by winter cold within its native range in Europe	Vanhanen et al. (2007), Régnière et al. (2009), Yasyukevich et al. (2015), Fält-Nardmann et al. (2018a), Streifel et al. (2019)
European pine sawfly	Range expansion northward in Europe	Virtanen et al. (1996)
Hemlock woolly adelgid	Model results are mixed regarding susceptibility to cold and potential range expansion northward in North America where it is invasive	Dukes et al. (2009), Fitzpatrick et al. (2012), McAvoy et al. (2017)
European spruce beetle	Increased population success through reduced generation time in Europe and in North America where it could become invasive; drought and severe storms increase outbreak frequency; increased disturbance areas	Jönsson et al. (2009, 2011), Økland et al. (2015), Seidl and Rammer (2017), Bentz et al. (2019)
Mountain pine beetle	Range shift northward in North America; increased population success at high elevations due to reduction in generation time; range contraction of historical distribution in United States due to disruption of seasonality; range expansion southward in North America; high establishment potential in parts of Europe	Bentz et al. (2010, 2016, 2019), Safranyik et al. (2010), Weed et al. (2015), Buotte et al. (2016), Sidder et al. (2016), Cooke and Carroll (2017)
Nun moth	Range expansion northward and eastward in Europe	Vanhanen et al. (2007), Yasyukevich et al. (2015), Fält-Nardmann et al. (2018b)
Red turpentine beetle	Population success in predicted invaded areas of Southern Hemisphere; expansion of populations in China where it is invasive	He et al. (2015), Lantschner et al. (2017)
Redbay ambrosia beetle	Range expansion northward in North America where it is invasive	Formby et al. (2018)
Sirex woodwasp	Range expansion in areas of Australia, Brazil, and northeastern North America where it is invasive	Carnegie et al. (2006), Lantschner et al. (2014), Ireland et al. (2018)
Southern pine beetle	Range expansion northward in North America	Lesk et al. (2017)
Spruce beetle	Increased population success through reduced generation time in North America in areas with suitable host trees	Bentz et al. (2010), DeRose et al. (2013)

(Continued)

Table 5 (*Continued*)

Common name	Prediction(s)	References
Spruce budworm	Range expansion northward in parts of North America (limited by thermal energy for life cycle completion); range contraction in southern United States due to diapause disruption	Gray (2008), Candau and Fleming (2011), Régnière et al. (2012), Pureswaran et al. (2015)
Western spruce budworm	Range expansion northward; less population success at lower latitudes (except at high elevations) due to exhaustion of energy reserves and mortality of overwintering larvae	Régnière and Nealis (2019)

range distributions, can provide estimates in the near-term but they do not adequately describe the nonlinear and evolved nature of insect responses to temperature, often do not include year-round temperature extremes that can influence forest insect success, and can be hampered by poor data on pest population distributions (Régnière et al., 2012; Barredo et al., 2015; Thompson et al., 2017; Mech et al., 2018). To adequately predict and manage for future responses of forest insects to a rapidly changing climate, process-based models based on species-specific phenological responses are needed. Quantification of thermal effects on complex nonlinear relationships among forest insects, their host trees, and community associates, including predators and microbial symbionts, will also enhance predictions (Dukes et al., 2009; Addison et al., 2015; Schwartzberg et al., 2014).

2.2 Plasticity and adaptive potential in a changing climate

Variation within and among forest insect populations in thermally regulated traits can potentially provide adaptive capacity to changing climatic conditions. Phenotypic plasticity, an organism's ability to adjust its phenotype in response to changes in environmental conditions, allows for higher fitness than if the trait was fixed, and populations with sufficient plasticity may persist in a rapidly changing environment in the short term without adaptation (Gienapp et al., 2008). Plasticity in thermally regulated traits has been demonstrated in some forest insect species including the European pine sawfly (*Neodiprion sertifer*) (Veteli et al., 2005), nun moth (*Lymantria monacha*), autumnal moth (*Epirrita autumnata*), European gypsy moth (in native and invaded ranges) (Keena, 2016; Fält-Nardmann et al., 2016, 2017; Thompson et al., 2017), mountain pine beetle (Bentz et al., 2011; Bentz and Hansen, 2018; McManis et al., 2018), European spruce beetle (Schroeder and Dalin, 2017), and the emerald ash borer (*Agrilus planipennis*) (Sobek-Swant et al., 2012a) and hemlock woolly adelgid (*Adelges tsugae*) (Butin et al., 2005; Lombardo and Elkinton, 2017) in invaded areas of North America. There are limits to plasticity, however, and long-term persistence *in situ* will require adaptations that occur through genetic evolution. Empirical data showing heritable genetic adaptation to rapidly changing conditions, however, are rare in insect pests (Garnas, 2018). In the one documented case for forest insects, van Asch et al. (2013) showed a genetic shift in the nun moth, from 2000 to 2010, to hatch their eggs later in spring to ensure that feeding larvae remain synchronized with bud break of host oaks. Recent northward range expansion of mountain pine beetle in Canada has also been associated with signals of directional selection on metabolic and cellular processes that may allow increased tolerance to cold temperatures (Janes et al., 2014).

3 Advances in managing for resistance in native and invasive forest insect systems

Tree resistance to herbivory involves a variety of physical and chemical defenses that are under genetic and environmental control. Harnessing resistance is key for future management of both native and alien invasive insect species (Telford et al., 2015). A framework was recently developed that is based on an in-depth analysis of factors associated with recurring failures in the long-term management of tree-killing, alien invasive insects in forest environments (Showalter et al., 2018). The analyses revealed that because alien invasive insects encounter defense-free spaces in their new environments, the evolutionarily naïve host trees are generally incapable of mounting adequate resistance responses. Top-down control by both native and introduced natural enemies can also be hampered, at times limiting the efficacy of biological control. Even the most undefended host populations, however, almost always include individuals that are capable of resisting attack. Such resistance need not be absolute (immunity), but sufficient to ensure survival and reproduction of the target host. Given survival, either natural selection can act directionally upon the traits conferring resistance in the long run, or modern approaches can be developed, including tree improvement programs, that are increasingly capable of rapidly selecting and augmenting tree defenses (Sniezko and Koch, 2017). In the latter case, improved trees could then be used for resilient plantings.

The practicality of tree-for-planting improvement programs is becoming more evident, as selection of resistant trees can be accelerated by using non-destructive resistance screening techniques based on modern -omics approaches. For example, intraspecific variability in host resistance can be detected by using a metabolomics-derived approach that includes infrared or Raman spectroscopic fingerprinting of phenolic extracts. Analysis of the chemical fingerprints is performed by using chemometric methods, which may include statistical approaches such as soft independent modeling of class analogy, principal components analysis, or, more generally, artificial intelligence tools such as artificial neural networks and support vector machines. Such analytical techniques are simply meant to classify plants as resistant or susceptible without reference to a specific mechanism (Conrad et al., 2014; Conrad and Bonello, 2016; Villari et al., 2017). Genomics and transcriptomics approaches have also been shown to be very promising in this respect (Harper et al., 2016).

Interspecific variability can also be helpful in understanding mechanisms of resistance that could then be developed into phenotyping tools for specific alien pest species. For example, little host resistance (but greater than complete susceptibility) appears to be available against emerald ash borer in North American ash species (Koch et al., 2015). Comparisons between North American naïve ash species and co-evolved, resistant Asian species such as Manchurian ash (*Fraxinus mandschurica*) have highlighted differences in the way they respond to emerald ash borer that are beginning to point to specific mechanisms that could be used for phenotyping. However, since interspecific comparisons can be confounded by genetic variation in unrelated traits, it is important to control for phylogeny. In the case of the emerald ash borer, the most relevant comparison has been between Manchurian ash (resistant) and black ash (*F. nigra*), the latter a highly susceptible North American species that belongs to the same section of *Fraxinus* as Manchurian ash (Wallander, 2008). Rigsby et al. (2015, 2016) have shown that resistance in Manchurian ash is characterized by higher activities of enzymes that appear to function by oxidative 'activation' of constitutive phenolics into quinones, among other mechanisms. Thus, screening for quinone generation may become a non-destructive tool to select more resistant North American ash individuals.

As would be expected from an evolutionary perspective, host tree resistance has been observed more frequently in native than alien invasive forest insect systems. Perhaps the most successful case of harnessing natural genetic variation for native species management is spruce (*Picea*) resistance to the white pine weevil (*Pissodes strobi*), a phloem feeder in regeneration-age spruce and white pines in North America. Heritable genetic variation in resistance to the white pine weevil has been well documented for multiple spruce species in western Canada (Kiss and Yanchuk, 1991; King et al., 2011). Evolved traits in spruce to weevil herbivory include stone cells that act as a physical barrier to feeding larvae and can slow development, thereby enhancing the toxic impacts of resin terpenes (Alfaro et al., 2004; Whitehill et al., 2019). Several seed orchards with resistant spruce for use in reforestation plantations have been established (Alfaro et al., 2013).

Published by Burleigh Dodds Science Publishing Limited, 2020.

In other systems, genetic variability in resistance to native insects has been identified, although tree breeding programs have not yet been developed. Recent research suggests high levels of genetic variation in susceptibility to the large pine weevil (*Hylobius abietis*), a pest of regeneration-age conifers in Europe, among Norway spruce (*Picea abies*) in southern Sweden (Zas et al., 2017) and maritime pine (*P. pinaster*) across its range (López-Goldar et al., 2018). Interestingly, in the latter study, inducibility of host chemical defenses was a critical factor in predicting resistance among host populations, whereas constitutive levels were not. Naturally occurring resistance of white spruce (*Pi. glauca*) to spruce budworm has also been documented (Daoust et al., 2010), and linked to the content of several specialized metabolites in spruce foliage (Delvas et al., 2011). Differences in gene expression and associated constitutive levels of the compounds were observed between white spruce individuals from different geographic regions with different intensities of historical spruce budworm outbreaks, suggesting that the traits are under positive selection (Parent et al., 2017). *Intraspecific* genetic variability was found among lodgepole pine (*Pinus contorta*) genotypes that were either attacked or not attacked by mountain pine beetle (Yanchuk et al., 2008), and *interspecific* variability in resistance was recently demonstrated where mountain pine beetle-caused mortality of *P. flexilis* was extensive, but absent in co-occurring *P. longaeva* of similar size (Bentz et al., 2017). The quantity and composition of constitutive specialized metabolites of *P. longaeva* phloem may play a role in resistance to mountain pine beetle attack and offspring production (Eidson et al., 2017, 2018). These research results suggest that rapid, non-destructive spectrometric screening tools could be developed, with immediate benefits to tree breeding programs.

4 Advances in native bark beetle monitoring and management

Bark beetles are important disturbance agents in conifer forests of boreal and temperate ecosystems, and several notable species are capable of causing significant amounts of tree mortality (Table 1). Trees of all species, ages, and size classes may be colonized and killed, but each bark beetle species exhibits unique host preferences, life history traits, and impacts. In recent years, climate change has contributed to some bark beetle outbreaks and range expansions in North America and Europe due to shifts in temperature and precipitation that influence both the beetles and their hosts (Table 4) (Section 2.1).

Bark beetles are part of the natural ecology of conifer forest ecosystems, important to proper functioning as they regulate certain aspects of primary production, nutrient cycling, and ecological succession. To that end, some level of tree mortality is desirable and often results in a mosaic of forest structure and composition that increases resistance and resilience to other disturbances. However, this varies from the ecological, economic, and social impacts of outbreaks, which can be significant (Morris et al., 2017). As a result, substantial basic and applied research has been devoted to the development of tools and tactics for mitigating undesirable levels of tree mortality attributed to bark beetles. Direct control involves short-term tactics designed to address current infestations by manipulating beetle populations, and often includes the use of sanitation harvests, semiochemicals (i.e. chemicals released by one organism that elicit a response, usually behavioural, in another organism), insecticides, or a combination of these and other treatments. Indirect control is preventive, and designed to reduce the probability and severity of future infestations by manipulating stand, forest, and/or landscape conditions by reducing the number of susceptible hosts through thinning, prescribed burning, and altering tree age classes, structure, and composition (Fettig and Hilszczański, 2015). Below, we provide a basic understanding of current and evolving management strategies. We encourage the reader to delve deeper into the literature cited and review the case studies on mountain pine beetle (Section 7.2) and larger eight-toothed European spruce bark beetle (= European spruce beetle), *Ips typographus* (L.) (Section 7.4), the most significant species in North America and Europe, respectively.

4.1 Survey and detection

Several methods are available to determine the distribution and severity of bark beetle infestations ranging from trapping programs to detect and monitor populations, to simple ground-based surveys, to a broad array of aerial surveys focused on detection of dead and dying trees (Fettig and Hilszczański, 2015). The most common method in North America incorporates the use of aircraft equipped with digital sketch-mapping systems (Fig. 2). In addition to showing the observer's position on a digital map, sketch mapping allows for real-time acquisition of data at lower costs (often <$ 1 USD/ha) than other survey methods. However, flying presents unique risks and considerable variability has been observed in data sketched by different observers. To address these concerns, recent research has focused on the utility of remotely sensed data obtained from satellites (Wulder et al., 2012) as the number and types of remote-sensing instruments have increased and image processing capabilities have greatly improved (Abdullah et al., 2019). The use of drone aircraft equipped with high-definition sensors and photographic equipment offers a new platform for survey and detection. Working in the central and southern Sierra Nevada, USA, Koontz et al. (2018) used drones to study the influence of spatial variability in forest structure on tree mortality attributed to western pine beetle (*D. brevicomis*). It is likely

Figure 2 Digital sketch-mapping systems are commonly used during aerial surveys of forest insect impact in North America. Photo credit: D. Wittwer, Forest Health Protection, USDA Forest Service.

that drones will play a larger role in the survey and detection of bark beetle infestations in the future.

4.2 Risk and hazard rating

Several risk and hazard rating systems are available to identify stands or forests that foster initiation and/or spread of bark beetle infestations (Fettig and Hilszczański, 2015). Most bark beetle species capable of causing extensive levels of tree mortality exhibit a preference for larger-diameter trees growing in high-density stands with a high percentage of host type (Fettig et al., 2007), and therefore such variables serve as a foundation for many risk and hazard rating systems. The accuracy of risk and hazard rating systems are influenced by many factors, and their refinement remains an important research priority as climate change seems to be influencing these relationships (Morris et al., 2017).

4.3 Direct control

Bark beetles have been the focus of direct control dating back to the 1700s. For example, the Royal Society of Sciences at Göttingen, Germany, established an award to recognize the best proposal for management of European spruce beetle in the mid-eighteenth century. In short, a successful direct control program requires prompt and thorough applications of the most appropriate tactics. However, as with indirect control, implementation is often dictated by more practical concerns such as resource availability (e.g. budget, time, personnel, and equipment), market conditions, market conditions, as well as equipping, logistical, and social constraints (Mattor et al., 2019).

4.3.1 Acoustics

Bark beetles use acoustics in a variety of behaviors, including territoriality (Rudinsky et al., 1976), mate recognition (Rudinsky and Michael, 1973), and predator escape (Lewis and Cane, 1990). In recent years, a renewed interest in the study of bark beetle acoustics may lead to novel management tools (Hofstetter et al., 2019). For example, Aflitto and Hofstetter (2014) demonstrated in laboratory studies that attacks on logs by southern pine beetle (*D. frontalis*) and western pine beetle were reduced by certain acoustic signals.

4.3.2 Biological control

Natural enemies are important in regulating bark beetle populations and have potential utility in biological control programs. In China, mass rearing and release of *Rhizophagus grandis*, a predatory beetle native to Eurasia, has been implemented in response to the introduction of the red turpentine beetle (*D. valens*) (Yang et al., 2014). *Rhizophagus grandis* is also used for management of the great spruce beetle (*D. micans*) in Europe where recent research focuses on how phenological relationships between predator and prey may be impacted by climate change (Gent et al., 2017). Research from North America indicates that conservation and supplemental feeding of natural enemies may enhance biocontrol efforts. For example, supplemental feeding of parasitoids of the southern pine beetle with an artificial diet consisting largely of sucrose increased their longevity and fecundity in pine forests (Stephen and Browne, 2000). Synthetic formulations of entomopathogenic microorganisms, such as fungi, bacteria, and viruses, are being developed and evaluated. Efforts have focused on the fungus *Beauveria bassiana*, which under laboratory conditions has been demonstrated to cause high levels of mortality in several bark beetle species. However, most field experiments have produced less than satisfactory control (Barta et al., 2018). Tactics under development include contaminating beetles collected in traps and then releasing these individuals back into field populations to contaminate the pest population (Kreutz et al., 2010), and applying various suspensions of *B. bassiana* spores to the surfaces of felled and standing trees (Davis et al., 2018).

4.3.3 Sanitation

Sanitation, one of the oldest direct control tactics (Gmelin, 1787), involves identification of currently infested trees and subsequent felling and removal or treatment to destroy adults that could emerge from the tree and brood beneath the bark. Where it is economically feasible, trees may be harvested and transported to mills where broods will be killed during processing. Otherwise, felled trees are debarked, chipped, burned, or treated by solarization (i.e. placement of infested material in the direct sun). In the southeastern USA, a unique version of sanitation involves harvesting trees infested with southern pine beetle plus a buffer strip of uninfested trees to halt infestation (spot) growth (Swain and Remion, 1981). Research on sanitation focuses on combining sanitation with other direct control methods (e.g. semiochemicals) to increase reductions in levels of tree mortality attributed to bark beetles.

4.3.4 Insecticides

Hundreds of thousands of high-value trees in campground and urban areas may be treated with insecticides during bark beetle outbreaks in the United States, yet similar uses of insecticides are widely restricted in most European countries. Most treatments involve topical sprays of contact insecticides where it is important that all parts of the tree that are likely to be attacked are adequately protected (Fettig et al., 2013b). Recent advances in systemic insecticides have led to the development of injection systems that push adequate volumes of product (i.e. generally less than several hundred milliliters for even large trees) into the small vesicles of the sapwood (Grosman et al., 2010). Following injection, the product is transported throughout the tree to the phloem where bark beetles feed. Compared to topical sprays, tree injections represent essentially closed systems that eliminate drift and reduce non-target effects and applicator exposure, but proper timing of applications is critical to allow for full distribution of the active ingredient within the tree prior to the tree being challenged by bark beetles (Fettig et al., 2014a, 2017). Research focuses on gaining a better understanding of the timing required in different systems (e.g. in high-elevation forests), and on combining systemic insecticides with fungicides to increase levels and duration of tree protection (Fettig et al., 2013b, 2014a).

4.3.5 Semiochemicals

The primary semiochemicals associated with the most aggressive bark beetle species have been isolated and identified, and combined with an integrated understanding of their context in the forest environment have led to several direct control methods (Seybold et al., 2018). For example, attractants such as aggregation pheromones and host kairomones are used to detect and monitor bark beetles, and to predict levels of tree mortality attributed to them. Attractants are also used in traps to collect and remove beetles from the forest (termed mass trapping), and placed on individual trees (termed trap trees) to induce attack prior to harvesting and sanitation. Repellents (inhibitors), such as the anti-aggregation pheromones verbenone and 3-methylcyclohex-2-en-l-one (MCH) and nonhost volatiles, are used to protect individual trees and forest stands (Zhang and Schlyter, 2004). Several formulations of anti-aggregation pheromones are commercially available (Seybold et al., 2018), and development of more effective blends (e.g. with nonhost volatiles) and formulations remains an important research focus (Fig. 3). In recent years, significant advances have also been made concerning the molecular biology and biochemistry of pheromone production in bark beetles, the synthesis of semiochemicals in the laboratory, the deployment of semiochemicals in the field, and the fate of semiochemicals once released into the active airspace of forests.

4.4 Indirect control

Numerous studies have demonstrated the effectiveness of reducing stand densities to increase resistance of trees to several bark beetle species (e.g. Wermelinger, 2004; Whitehead et al., 2006; Fettig et al., 2007, 2014b; but see Six et al., 2014). For example, Negrón et al. (2018) quantified the impacts of thinning implemented 2 years prior to a mountain pine beetle outbreak in the Black Hills, USA, and reported that the percentage of ponderosa pine (*P. ponderosa*) killed by mountain pine beetle during the outbreak was 34% in unthinned stands and 4% in thinned stands. In Europe, Heidger and Lieutier (2002) recommended thinning of spruce in order to reduce competition among trees, thereby increasing availability of water, nutrients, and sunlight allowing for enhanced growth and defense of residual trees against European spruce beetle. Thinning also has important influences on microclimate and tree spacing that affect resistance to bark beetles.

Figure 3 A relatively new bark beetle repellent for protection of individual conifers is SPLAT® Verb, a flowable and biodegradable formulation of the anti-aggregation pheromone verbenone that allows the user to adjust the size of each release point (dollop) in the field (Fettig et al., 2015). The inert ingredients are certified as food-safe by the US Environmental Protection Agency (EPA), and the product is classified as organic by the US Department of Agriculture. Other semiochemical-based tools for management of bark beetles with the SPLAT® matrix are under development. In this figure, multiple dollops of SPLAT® Verb have been applied to each aspect of the tree bole to study the effect of release rates on tree protection. Photo credit: R. Progar, Washington Office, USDA Forest Service.

For example, thinning may cause changes in wind speeds and direction within stands that cause turbulences that disrupt pheromone plumes used for recruiting conspecifics to the target tree during initial phases of host colonization (Thistle et al., 2004). This negatively affects host-finding success. Recent research indicates that thinning is effective for increasing resistance to several bark beetle species even during 1000-year drought events (Negrón et al., 2009; Santos and Whitman, 2010; Fettig et al., 2019; Restaino et al., 2019). However, given projected increases in drought stress (Vose et al., 2018), tree-density (stocking) thresholds are being reevaluated and lowered to maintain adequate levels of resistance to bark beetles during severe droughts (e.g. in the southwestern United States). Prescribed fire may also be used to manage stand densities in some forests, and while short-term (e.g. 1-5 years) increases in bark beetle-caused tree mortality are likely (due to stress induced by sublethal heating of plant tissues) in the longer term burned areas benefit from the positive impacts of prescribed fire on growing space (Fettig and McKelvey, 2014).

5 Advances in defoliator monitoring and management

The majority of defoliator pest species of boreal and temperate forests are within the order Lepidoptera (Table 2). Population regulation of defoliators occurs naturally *via* bottom-up (host tree defense) and top-down (natural enemy) controls. At low-to-moderate levels of defoliation, tree growth may decline, but most trees survive. During population outbreaks, large-scale tree mortality can occur following cumulative defoliation over several years. Protecting forests from outbreaks of defoliators is a complex, expensive, and often controversial process, especially when it involves harvesting of trees and aerial spraying of insecticides. Aerial application of insecticides began in the late nineteenth century, but due to increasing environmental concerns has been largely replaced by biological control.

5.1 Direct control

5.1.1 Microbial biological control

Microbial biological control, which includes entomopathogens such as viruses, bacteria, and fungi, are commonly used in management of forest defoliators (see Sections 7.3 and 7.5). In North America, the *Lymantria dispar* nucleopolyhedrosis virus (LdNPV) is the most documented factor causing the collapse of European gypsy moth populations (McManus and Csoka, 2007; Hajek and van Nouhuys, 2016). With modern molecular technology, complete genome sequences are known for several forest insect viruses including NPV associated with several budworm species (*Choristoneura* spp.), white marked tussock moth (*Orgyia leucostigma),* European gypsy moth (Harrison et al., 2016), and sawflies (*Neodiprion* spp.) (van Frankenhuyzen et al., 2016). Application of sequence homologies has allowed a specific insect species to be targeted, and thereby facilitated approval of viruses in control strategies used by regulatory agencies (Lapointe et al., 2012).

The first commercial product of a bacterium, *Bacillus thuringiensis,* was produced in California in 1957. Now commonly known as Bt, the strain *B. thuringiensis* var. *kurstaki* (*Btk*) has since been used as an alternative for chemical insecticides in large-scale aerial spray operations to control multiple defoliator species (van Frankenhuyzen et al., 2016; Fuentealba et al., 2019). A ubiquitous microsporidian pathogen of spruce budworm, *Nosema fumiferanae*, was evaluated for microbial control in field trials during a spruce budworm outbreak in eastern Canada in the 1970s. However, recent studies indicate that it may act in concert with other density-dependent factors, and that direct mortality due to this pathogen is limited (Eveleigh et al., 2012). The use of molecular techniques will increasingly facilitate the development of novel microbial-based pest management options.

5.1.2 Natural enemy (insect) biologica control

Defoliator outbreaks occur when control by natural enemies becomes ineffective, allowing relatively uncontrolled population growth. Natural control of forest defoliators occurs largely by parasitoids (Nealis, 1991), primarily wasps and flies that deposit eggs on, in, or near specific immature life stages of the defoliator (i.e. eggs, larvae, or pupae). Parasitoid larvae later hatch and consume the defoliator host. Classical biological control takes advantage of the role of natural enemies, including parasitoids, in regulating population growth. Biological control of defoliators has been implemented for at least a century, with varying success (MacQuarrie et al., 2016; Kenis et al., 2017), by rearing and releasing natural enemies into outbreak areas. Most biological control programs in Europe and North America have targeted non-native pest species by introducing a non-native biocontrol agent from the native range of the non-native pest. In Europe, around 750 parasitoids and predators have been introduced against lepidopteran pests of woody plants since 1870, of which only 2% are considered successful (Kenis et al., 2017). In North America, introduced and established parasitoids of European gypsy moth can cause significant mortality of larvae and pupae (McManus and Csoka, 2007; Fuester et al., 2014).

In many defoliator species, regulation by natural enemies determines outbreak intensity and duration. For example, specialist parasitoids can have a rapid numerical response and increased capacity to control outbreak populations of the forest tent caterpillar (*Malacosoma disstria*) and hemlock looper (*Lambdina fiscellaria*), resulting in frequent outbreaks of short duration (Hébert et al., 2001; Roland, 2005). In contrast, the parasitoid community affecting the spruce budworm is dominated by generalist species whose slow numerical response (Eveleigh et al., 2007) creates outbreak cycles of relatively longer period and duration (Régnière and Nealis, 2007). The abundance and efficacy of parasitoids varies with landscape structure (Roland and Taylor, 1997), and increased forest diversity has been associated with slower expansion rates of spruce budworm outbreaks and lower outbreak intensity (Robert et al., 2018). Forest fragmentation and losses in diversity can therefore influence outbreak duration and spread of defoliators through effects on their natural enemies (Roland, 1993).

5.1.3 Semiochemicals

The use of semiochemicals for management of defoliators includes pheromones to attract mates or conspecifics and kairomones (i.e. plant volatile cues) (Evenden and Silk, 2016). The most common uses of sex pheromones are for detection, monitoring (Schmidt et al., 2003), population suppression (reducing populations to mitigate damage from defoliation), implementing barrier zones to slow spread from infested areas (e.g. for gypsy moth 'Slow The Spread' (STS), see Section 7.5.2), and eradication of isolated infestations in uninfested areas. Mass trapping uses pheromones to attract insects to traps where the pests are then killed with an insecticide. Mating disruption reduces mating success by disorienting males through saturation of the environment with pheromone (Witzgall et al., 2010). (+)-Disparlure attracts male European gypsy moths, and baited 'delta traps' have been deployed in western Canada in a grid across landscapes to detect and eradicate moth populations that have been repeatedly intercepted but not successfully established (Nealis, 2009). 'Disrupt Micro-flake spruce budworm' (Hercon Environmental, Emigsville, Pennsylvania, USA) was aerially applied to stands infested by spruce budworm in Canada where it reduced mate-finding by males, but there was no decrease in the number of egg masses laid (Rhainds et al., 2012; Evenden and Silk, 2016). Dispersal of female moths into treated areas is thought to limit the effectiveness of mating disruption in the spruce budworm. In contrast, mating disruption with sex-attractant pheromones has been a key component of the European gypsy moth STS program in the United States (Onufrieva et al., 2019) (Section 7.5.2). These differences highlight how targeting isolated populations with mating disruption can result in greater efficacy than when targeting widely established outbreak populations.

5.2 Indirect control

Bottom-up control of defoliators can be accomplished by creating healthy, vigorous forest stands *via* silvicultural practices. Stressed trees in forest plantations are susceptible to defoliators and can facilitate buildup of populations that can trigger a larger outbreak (Björkman et al., 2015; Muzika, 2017).

Published by Burleigh Dodds Science Publishing Limited, 2020.

Maintaining high tree species diversity can decrease buildup of specialist defoliator populations by conferring 'associational resistance' to host species of the pest when it occurs among nonhosts (Jactel et al., 2017). There is evidence that pre-commercial thinning to increase black spruce proportions in mixed fir-spruce stands, or increasing hardwood content in the landscape could reduce overall tree mortality during spruce budworm outbreaks (Campbell et al., 2008; Pothier et al., 2012). Fragmenting the forest landscape by planting nonhost tree species can disrupt host-finding behavior of ovipositing females by making suitable hosts relatively less apparent. In generalist defoliators, however, the response to tree diversity can be variable (Moreira et al., 2016). Tree species diversity can also increase the diversity and abundance of generalist parasitoids that would serve as natural controls and potentially decrease population growth rates (Roland and Taylor, 1997). Maintaining stand age diversity would also decrease density of mature trees that may be susceptible to defoliation. Finally, removal of susceptible trees by logging and infested branches by pruning in the wake of an outbreak can mitigate the impacts of defoliation. Low population densities of tent caterpillars (*Malacosoma* spp.) in North America and processionary moths (*Thaumetopoea* spp.) in Europe and North Africa can be controlled by mechanically removing larval colonies (Ciesla, 2011).

6 Advances in invasive species monitoring and management

Non-native invasive forest insects have threatened and damaged trees in northern boreal and temperate forests for over 150 years (Liebhold et al., 1995; Niemelä and Mattson, 1996), and, more recently, they have become a major focus of forest insect management (Langor et al., 2009; Aukema et al., 2010; Liebhold et al., 2017). Native invasive species that have been moved or have dispersed into new ecological contexts within the geographical borders of countries have also been vexing, both biologically and politically (Table 3, e.g. Baranchikov et al., 2010; Seybold et al., 2019). Many of these invasive species have had only minor, localized, or regional impacts on forest or shade trees, but some have become major pests with landscape-level impact (Table 3, Tobin and Liebhold, 2011; Herms and McCullough, 2014). In an assessment of more than 450 invasive forest insect species that have established populations in the United States, Aukema et al. (2010) considered 14% (62 species) as 'high impact' species, that is, of regulatory significance or having caused notable damage to forests or urban forest trees. The US Department of Agriculture (USDA) Forest Service has also recently completed a comprehensive assessment of invasive species with a focus on those organisms that interact with forest and rangeland ecosystems (Poland et al., 2019). The assessment covers the full range of issues from risk assessment, monitoring/detection, impact, and management to economics and sociological issues. It also provides summaries that highlight invasive species of concern by US region. In this section, we will focus on the most recent advances in detection, monitoring, and management of invasive forest insect species.

6.1 Detection and monitoring

Traditionally, the detection of invasive forest insects was a matter of chance encounters through hand collection or observation of sudden and surprising tree damage patterns (Coleman and Seybold, 2008). Taxonomic specialists kept records and maintained primary awareness of native and adventive distributions of potential pest taxa (Wood and Bright, 1992). More recently, systematic approaches to detection have emerged by targeting ports of entry, urban parks and arboreta, green waste facilities, agricultural habitats, and similar areas through inspection and trapping with generic and specific semiochemical lures (Seybold et al., 2016; Rabaglia et al., 2019). There is also growing recognition that trees in urban and agricultural habitats can provide landscape-level bridging for forest pests into forest habitats (Rossi et al., 2016). Diverse collections of plants organized into geographically dispersed and formalized networks can also provide powerful opportunities for early detection of invading insects that impact trees (Barham, 2016). The invasion process has been preempted still further by focusing on 'overseas' surveillance where, for example, United States and Canadian forest entomologists have long pursued a cooperative assistance program with their foreign counterparts in and around ports of entry in the Russian Far East, Japan, Korea, and China to detect and manage potential lepidopteran invaders (USDA, 1993; Humble et al., 2013; Freyman, 2015; USDA APHIS, 2016). Recently, this concept has been extended to Asian-sourced ambrosia beetles and their pathogenic fungi, which have emerged in the early twenty-first century as preeminent invasive pests of trees (Hulcr and Dunn, 2011; Hulcr et al., 2017). The development of molecular techniques based on multiple genes or even full genomes has facilitated confirmatory identifications of species, subspecies, strains, and biotypes to provide an informative end point to the detection process (EPPO, 2016; Thomas et al., 2016; Bilodeau et al., 2019; Roe et al., 2019).

Risk assessment and pathway analysis have also received greater attention in the planning and allocation of limited resources to address invasive species monitoring and detection (Venette et al., 2010; Fuentealba et al., 2013; Susaeta et al., 2016; Hudgins et al., 2017). For example, Gray (2017) described an analysis of the risk of the invasion pathway of the Asian gypsy moth (*Lymantria dispar asiatica*) into North America through an evaluation of its 'stowaway' pathway on containers or cargo on ships from regulated (Asian) ports to protected (North American) ports. The analysis involved the consideration of the propagule size on ships; the probability of propagule release in the protected port; and the survival and establishment of propagules in that port. In this instance, the propagules are newly hatched larvae. A gypsy moth phenology model specific for the location of the protected port was useful for providing the pathway specifics for the three criteria.

6.2 Management

Management strategies for invasive forest insects rely on tactics described in the earlier sections on bark beetles (Section 4) and defoliators (Section 5) with the addition of the potential for preventing arrival of incipient populations through rigorous quarantine and inspection programs, as well as local or complete eradication of nascent populations. When prevention or eradication of an invading population is not possible, retarding the spread rate throughout the new range may have economic and ecological benefits (see Section 7.1.2: SLow Ash Mortality (SLAM) for the invasive emerald ash borer and 7.5.2: Slow The Spread (STS) for the European gypsy moth as two North American examples). Generally, extensive efforts have been made to carry out classical biological control programs to exert some level of natural population regulation of an invasive forest insect (e.g. Duan et al., 2014). Similarly, cultural practices involving re-planting with resistant cultivars of host species or closely related species have been investigated and attempted (Rebek et al., 2008; Villari et al., 2016) (see Section 3). After establishing the impact of an invasive forest insect species on the adventive forest resources (Muzika and Liebhold, 1999; Eyles et al., 2011; Coleman et al., 2012), there has been increasing attention and effort to apply silvicultural methods to manage well-established populations of invasive species (Dodds et al., 2014; Muzika, 2017). Thus, the full range of integrated pest management techniques can in theory be applied to invasive forest insects once they have become established in a new habitat.

7 Case studies

7.1 Emerald ash borer

Since its discovery in southeastern Michigan, USA, in 2002, the emerald ash borer (see Table 3) has killed hundreds of millions of ash trees with devastating economic and ecological impacts and threatens the entire North American ash resource (Poland and McCullough, 2006; Gandhi and Herms, 2010; Kovacs et al., 2010; Flower et al., 2013; EAB Info, 2019). Early efforts to eradicate isolated infestations were unsuccessful; therefore, considerable research has been conducted to develop effective detection and management tools in North America.

7.1.1 Early detection, monitoring, and management

Early detection of low-density infestations is notoriously difficult because of emerald ash borer's cryptic oviposition and feeding behavior under the bark and lack of external signs or symptoms until trees are severely damaged and adults have emerged (Poland and McCullough, 2006). Adults are attracted to stressed or girdled ash trees (McCullough et al., 2009a,b), leaf and bark volatiles including *cis*-3-hexenol (de Groot et al., 2008) and sesquiterpenes found in Manuka oil (Crook et al., 2008), and shades of green or purple color (Francese et al., 2010a). A female-produced pheromone, *cis*-lactone (a macrolide), may also enhance close range attraction (Ryall et al., 2012). The most promising trap designs for early detection of emerald ash borer include green funnel traps (Francese et al., 2011), green or purple sticky prism traps hung in the ash canopy (Francese et al., 2010b), or green or purple double-decker traps (Poland et al., 2011; Poland and McCullough, 2014). Early identification of infestations with traps allows for rapid implementation of management tactics.

Current management strategies for emerald ash borer in North America have shifted toward integrated area-wide programs designed to reduce population growth in infested areas and slow its spread to non-infested areas. Integrated management programs include public outreach (Mercader et al., 2013); surveys by using artificial traps and girdled trap trees; quarantines and other regulatory activities that reduce risk of human transport of emerald ash borer-infested materials (USDA APHIS, 2019); removal and destruction of infested trees (McCullough et al., 2015; Mercader et al., 2015, 2016); application of systemic insecticides to protect trees (Smitley et al., 2010; McCullough et al., 2011); and release of natural enemies for biological control (Duan et al., 2014; Bauer et al., 2015; Jennings et al., 2016).

Insecticide treatments, such as trunk injection of systemic insecticides (emamectin benzoate, azadirachtin), basal trunk sprays (dinotefuran), or soil applications (imidacloprid, dinotefuran), are effective at protecting high-value landscape trees in urban areas (McKenzie et al., 2010; Smitley et al., 2010; McCullough et al., 2011), although costs and environmental concerns prevent widespread use in forested areas. Therefore, biological control is being evaluated for long-term sustained management of emerald ash borer in natural forests (Bauer et al., 2015). Two of three introduced Asian parasitoids (*Tetrastichus planipennisi* and *Oobius agrili*) have established stable populations in several US states (Duan et al., 2018). These species, along with some native natural enemies, have substantially reduced emerald ash borer population growth in some locations (Duan et al., 2014). A fourth parasitoid native to Russia (*Spathius galinae*) was approved for release in 2015 (USDA APHIS, 2015). Research is also being conducted to develop hosts that are resistant to emerald ash borer for use in restoration of heavily impacted ecosystems (Koch et al., 2015; see Section 3).

7.1.2 SLow Ash Mortality (SLAM)

Multiple agencies in the United States including the USDA Forest Service, USDA APHIS, Michigan Department of Natural Resources, Michigan Department of Agriculture and Rural Development, Michigan State University, and Michigan Technological University partnered to conduct a large pilot study to evaluate the SLAM integrated strategy for managing the emerald ash borer (Poland and McCullough, 2010; McCullough and Mercader, 2012). Though a very small proportion of ash trees in the >390 km^2 project area were treated with insecticides or girdled, both tactics reduced emerald ash borer densities and protected ash trees in areas surrounding the treatments (Mercader et al., 2015). Recently, area-wide management approaches integrating biological control and selective use of insecticides have been initiated. Integrating these components may yield additive or synergistic effects (McCullough and Mercader, 2012) as insecticides impact newly hatched neonate larvae and adults that both feed on leaves (Smitley et al., 2010; McCullough et al., 2011), whereas natural enemies parasitize eggs or later-instar larvae (Bauer et al., 2015). Landscape-level integrated management strategies such as SLAM and biological control, insecticide treatments in urban areas, collection and preservation of ash seed, and development of more resistant ash (Section 3) offer hope for the protection of ash and persistence of the genus at some level in urban environments and forests of North America.

7.2 Mountain pine beetle

Mountain pine beetle is the most significant forest insect pest in North America, and colonizes multiple pine species (Negrón and Fettig, 2014; Table 1). The geographic distribution generally reflects the range of its primary hosts, although lodgepole pine extends further to the north in Canada and ponderosa pine and other pines extend further to the south in Mexico than where mountain pine beetle populations currently exist (Bentz et al., 2019). Since 2000, ~10.3 million ha have been impacted by mountain pine beetle in the United States alone (Berner et al., 2017), which represents almost half of the total area impacted by all bark beetles species combined in the western United States during this period.

Historically, the range of mountain pine beetle was restricted by climatic conditions unfavorable to brood development and survival, despite the availability of host trees, but the pest distribution is expanding northward due to climate change (Table 5) and other factors including lower defenses in novel hosts (Erbilgin, 2019). By the end of the twenty-first century, thermal suitability for mountain pine beetle population success is projected to be high at the most northern extent of pines in Canada, although portions of the historical range are projected to become unsuitable due to excessive warming that disrupts overwintering and adult emergence timing (Bentz et al., 2010, 2019). Models also project that thermal suitability for mountain pine beetle in pine forests further south of the current range, including Mexico and the southwestern United States, will increase throughout the century. Mountain pine beetle is the second most intensively studied bark beetle after European spruce beetle (see Section 7.4). A century of research has produced thousands of scientific papers and a wealth of knowledge applicable to the management of this species (Table 6), much of which is captured in several syntheses (e.g. Berryman et al., 1978; Safranyik and Wilson, 2006; Negrón and Fettig, 2014). Researchers continue to develop valuable knowledge and tools, of which we highlight a few of many recent examples below.

7.2.1 Fuels and fire behavior

During the previous decade, substantial research on the effects of mountain pine beetle outbreaks on forest fuels and surface and crown fire behaviors has revealed that mountain pine beetle and wildfires are principal drivers of change in western North American forests. Much of the knowledge generated has been synthesized by Jenkins et al. (2008, 2014), Hicke et al. (2012), and Kane et al. (2017). Practical implications relevant to firefighter safety are discussed in Jenkins et al. (2012, 2014).

Immediately following tree colonization, changes occur in needle moisture content and chemistry that influence flammability (Page et al., 2012). As time progresses (years to decades following an outbreak), significant increases occur in the accumulations of large coarse woody surface fuels (100- and 1000-hr fuels) (Jenkins et al., 2014). As such, several attempts have been made to predict fire behavior in mountain pine beetle-affected stands by using existing fire behavior models. Generally, predicted surface fire rates of spread and fire line intensities are higher during (e.g. initial 1-5 years) and then again decades after an outbreak. The literature is more mixed concerning outbreak effects on crown fire behavior, and confounded by the heterogeneity of crown fuels in beetle-impacted forests that violates the fundamental assumptions of homogeneity in the application of fire behavior models (Cruz and Alexander, 2010). Regional-scale studies have reported negligible effects on fire severity and scale (e.g. Hart et al., 2015; Meigs et al., 2015), especially under extreme fire weather conditions. Differences in the severity and distribution of outbreaks, time since outbreak (i.e. the 'bark beetle rotation'; Jenkins et al., 2008), dominant fire regime, and the retrospective nature of most studies challenge our understanding of these relationships.

7.2.2 Semiochemical-based tools

Research concerning semiochemical-based tools and tactics for mountain pine beetle has progressed steadily since the late 1960s, resulting in tactics

Table 6 Monitoring and management tools for mountain pine beetle

Detection and survey	Many methods are available, but aerial detection surveys using trained observers with digital sketch-mapping equipment remains most relevant (Fig. 2). Recent advances have been made with remotely sensed data from satellites, but there are limitations to their applicability (Wulder et al., 2012).
Risk and hazard rating systems	Several risk and hazard rating systems are available. The most relevant is that of Shore and Safranyik (1992) for use in lodgepole pine. Susceptibility is calculated based on four factors: (1) percentage of susceptible basal area (trees ≥15 cm dbh); (2) average stand age of dominant and co-dominant trees; (3) stand density of all trees ≥7.5 cm dbh; and (4) the geographic location of the stand in terms of latitude, longitude, and elevation. The proximity and size of local mountain pine beetle populations are added to compute an overall index. Several systems are available for use in other host types (Fettig et al., 2014c).
Insecticides	The use of insecticides for tree protection involves topical sprays and systemic insecticides injected directly into the tree (Fettig et al., 2013b). Remedial applications to kill adults and brood within currently infested hosts are no longer used.
Semiochemicals	The use of semiochemicals includes applications of aggregation pheromones and co-attractants (e.g. host volatiles) for survey and detection and trap-out, trap-tree, or push-pull tactics; and applications of repellents (inhibitors) for protection of individual trees and forest stands (Progar et al., 2014; Seybold et al., 2018). Verbenone is the most widely used repellent semiochemical.
Sanitation	In some cases, emphasis is placed on removal of newly infested trees to reduce the quantity of attractive semiochemicals (i.e. aggregation pheromones and host volatiles) released into the forest stand, but this is rare due to complications regarding the identification of newly attacked trees and the level of responsiveness required in their removal. Sanitation combined with the use of verbenone is recommended for mitigating small infestations in lodgepole pine forests (Progar et al., 2014).
Thinning	Thinning has long been advocated as a preventive measure. Common tactics include thinning from above or diameter-limit thinning; thinning from below; removal of trees with thick phloem; and increasing residual tree spacing (Fettig et al., 2014c). Thinning is likely most effective when implemented several years (to decades) prior to an infestation, and when appropriate stand densities are maintained over time (but see Negrón et al., 2018).
Increasing landscape heterogeneity	Increasing heterogeneity (e.g. of age, size, and tree species compositions) increases resistance of forested landscapes to mountain pine beetle (Fettig et al., 2014c).
Reducing the rate of atmospheric warming	Forests are increasingly vulnerable to mountain pine beetle as a result of the direct and indirect effects of climate change (Bentz et al., 2019).

for monitoring and detection, and for reducing levels of tree mortality at various spatial scales. Much of this work has been synthesized by Progar et al. (2014) and Seybold et al. (2018). Recent advances include development and commercialization of verbenone-releasing flakes for aerial treatment of large (e.g. >100 ha) areas (Gillette et al., 2009, 2012), and of a flowable and biodegradable formulation that is hand applied (SPLAT® Verb; Fettig et al., 2015b) (Fig. 3). After researchers demonstrated that combinations of green leaf volatiles and angiosperm bark volatiles could disrupt attraction of mountain pine beetle to baited trees (see Seybold et al., 2018), nonhost volatiles became the focus of numerous studies in hopes of increasing levels of tree protection over those from application of verbenone alone. Perhaps most notable, Fettig et al. (2012a,b) developed Verbenone Plus [acetophenone, (*E*)-2-hexen-1-ol + (*Z*)-2-hexen-1-ol, and (–)-verbenone], which has been demonstrated to be effective for tree protection in several studies in western Canada and the western United States, but has not yet been commercialized. An obstruction to the use of semiochemical-based tools for tree protection is the high cost of production, but recent advances in synthesis have lowered costs (Chou and Keasling, 2012).

7.2.3 Insecticides

Fettig et al. (2013b) discuss the efficacy, residual activity, and environmental safety of insecticides used to protect high-value trees near homes and in campgrounds from mountain pine beetle attack. Recent advances include the development and commercialization of several contact insecticides and alternative timings of associated treatments (Fettig et al., 2006a,b, 2015a, 2018), and of systemic insecticides applied through tree injection systems (Grosman et al., 2010; Fettig et al., 2014a,b). Other research has focused on the environmental fate of insecticides resulting from conventional spray applications, where the amount of drift that occurs was found to pose little threat to adjacent aquatic environments (Fettig et al., 2013b).

7.3 Spruce budworm

Regional outbreaks of spruce budworm (Table 2) have occurred about every 35 years in Canada and the northeastern United States on balsam fir (*Abies balsamea*) and spruce, leading to development of integrated management strategies. Planting of nonhosts, including pines, rather than host species and increasing the proportion of hardwoods on the landscape can reduce population buildup (Campbell et al., 2008; Pothier et al., 2012).

7.3.1 Spruce Budworm Decision Support System

The Spruce Budworm Decision Support System (SBWDSS) relies on forest inventory, stand yield, and growth models, as well as harvest scheduling information to estimate losses in timber supply and to facilitate forest management operations (MacLean et al., 2002). In addition to preventive silviculture operations (Pothier et al., 2012), mature, high-value stands are sprayed with biological control products (see Section 5.1) above the canopy with aircraft. The most frequently used is *Btk* (see Section 5.1.1), a microbial insecticide that causes mortality by rupturing the midgut of feeding larvae, quickly halting feeding, and protecting trees from further defoliation (Fuentealba et al., 2019).

7.3.2 Early intervention strategy

Following the last outbreak of spruce budworm in eastern Canada (ca. 1960–1992) that resulted in losses of 382 million m^3 of timber between 1977 and 1987 alone (Sterner and Davidson, 1982), the implementation of an "early intervention strategy" is being tested during a current outbreak that began in Quebec, Canada, in 2006. Spruce budworm is a highly mobile species, and control actions aim to suppress populations prior to large-scale defoliation events, and early enough to prevent moth dispersal into neighboring uninfested forests. The early intervention strategy is based on principles of population dynamics and Allee effects that can diminish population growth rates. Allee effects cause a decline in per capita population growth when population density decreases below a critical threshold and can limit establishment when sufficient mates are not available (Liebhold and Tobin, 2008). At low population densities, spruce budworm is subjected to strong predator-driven and mate-finding Allee effects (Régnière et al., 2013). Unless populations are subsidized adequately by eggs from immigrating moths, decreasing growth rates occur. A main goal of the early intervention strategy is to reduce the density of newly detected outbreak spots below thresholds for mating success. Several products are used to accomplish this including *Btk* and aerial application of an insect growth regulator tebufenozide (Mimic®) and a pheromone-based mating disruptant (Disrupt Micro-Flake SBW®). Current tests began in 2014 in New Brunswick, which is geographically a potential sink for dispersing moths from the outbreak in Quebec.

Current research focuses on evaluating the ability of treatments to (1) reduce spruce budworm populations to levels below Allee thresholds where negative density-dependence occurs, (2) evaluate the rate at which immigrating moths re-invade areas where populations have been suppressed, and (3) evaluate the long-term response of natural enemies to the suppressed spruce budworm population (SBW EIS Phase 2 Project Proposal 2017). Early results demonstrate that spruce budworm populations in treated "hot-spots" have lower larval survival rates compared to control stands (Martel et al., 2018). If successfully implemented, this strategy, which integrates basic science with applied science as well as emerging technologies, can mitigate timber losses and the need for costly control measures during an outbreak.

7.4 European spruce beetle

The European spruce beetle is the most important forest insect in Europe, and colonizes predominantly Norway spruce (Table 1). Outbreaks are often incited by other disturbances, such as windstorms, which provide an abundance of weakened host material that fosters rapid increases in population densities (Marini et al., 2017). Due to vast forest exploitations in the past and current wood demands, Norway spruce is now widely distributed outside its native range (Kölling et al., 2009), and is frequently found on unsuitable sites in Central Europe where it is highly susceptible to European spruce beetle.

In the last three decades across eight European countries, the amount of timber damaged by European spruce beetle outbreaks was driven by increasing summer rainfall deficit (drought), warm temperatures, and the density of storm-felled trees from the previous year (Marini et al., 2017). Warming temperatures are increasing the area of spruce habitat that supports two rather than one generation per year (Netherer et al., 2015) and a higher number of sister broods (Davídková and Doležal, 2017), both of which can benefit population growth. Models have been developed to predict how changing temperatures will influence these traits (Baier et al., 2007; Jönsson et al., 2009, 2011; Siedl and Rammer, 2017) (Table 5; Section 2.1). In one case study, model projections suggest that these trends will continue throughout the twenty-first century with an ~30% increase in European *Picea* habitat supporting a second generation (Bentz et al., 2019). Model results also project high thermal suitability for European spruce beetle population growth in potentially invadable *Picea* habitat in North America throughout this century, highlighting the importance of continued monitoring at North American ports (see Section 6.1).

Several centuries of research have produced thousands of scientific papers and a wealth of knowledge applicable to the management of European spruce beetle (Table 7), much of which has most recently been synthesized by Wermelinger (2004) and Fettig and Hilszczański (2015). Since publication of the first forestry text that addressed bark beetles (Ratzeburg, 1839), control methods have been developed systematically and employed regularly to reduce tree losses attributable to this species. During outbreaks, thousands of trees are attacked and killed within weeks prompting large and well-coordinated management efforts by several countries (e.g., Table 8). Researchers continue to develop valuable knowledge and tools, of which we highlight a few of many recent examples below.

7.4.1 Biocontrol

To date, biological control has not been formally implemented for European spruce beetle. However, some recent efforts concerning development of *B. bassiana* are considered to be promising. Research focuses on improving the virulence of formulations under field conditions, and on developing effective methods of delivery of *B. bassiana* into wild populations of European spruce beetle (Kreutz et al., 2010; Barta et al., 2018). Patents have been procured for a low-molecular polyethylene matrix that stabilizes *B. bassiana* spores and conidia, and for a trap that serves as an autoinoculation device for captured individuals.

7.4.2 Semiochemical-based tools

Research concerning semiochemical-based tools and tactics for management of European spruce beetle has progressed steadily since the 1970s, but despite this, they are largely limited to the use of attractants for survey and detection, mass trapping (trap-out), and trap-tree methods. This is somewhat of an artifact of the effectiveness and prevalence of sanitation for control of European spruce beetle throughout most of Europe (Tables 7 and 8). In recent decades, the development of behavioral repellents (inhibitors) has received some attention (Zhang and Schlyter, 2004); however, the most active blends contain several components in addition to verbenone (e.g., Schiebe et al., 2011), which are too costly for widespread use. As such, recent research has focused on refining these blends and reducing the number and cost of components required to impart tree protection. Unelius et al. (2014) reported that a blend consisting of verbenone, 1-hexanol, and *trans*-conophthorin is a cost-efficient repellent warranting more widespread use where sanitation is not feasible. In Germany, related research focuses on developing strategies that apply SPLAT® Verb to prevent the buildup of high population densities after disturbances such as windstorms.

7.5 European and Asian gypsy moths

Larvae of gypsy moths and related species (e.g., nun moth, pink gypsy moth, *L. mathura*, and other Asian species such as *L. albescens*, *L. postalba,* and *L. umbrosa*) are noted defoliators of forest trees (Tables 2, 3, and 5; Pogue and Schaefer, 2007; Tobin and Liebhold, 2011). The European gypsy moth, primarily a hardwood defoliator, was introduced and established in the eastern portion of the United States in the late 1860s (Liebhold et al., 1989). The Asian gypsy moth originates from temperate Asia (i.e., north of the Himalayas) ranging from the Ural Mountains east to China, Korea, the Russian Far East, and Japan, and there have been frequent introductions and attempted eradications of this

Table 7 Monitoring and management tools for European spruce beetle

Risk and hazard rating systems	Several attempts to develop risk and hazard rating systems have been made or are being developed (e.g., Netherer; Nopp-Mayr, 2005, Kautz et al., 2018). More sophisticated models, such as PHENIPS, provide a tool for hazard rating at local and regional scales (Baier et al., 2007).
Detection and survey	Infested trees are located *via* ground surveys executed by well-trained technicians. Initial attempts using unmanned aerial vehicles for survey and detection have failed (Ackermann et al., 2018). Johansson et al. (2019) suggest that dogs are able to detect infested spruce from the first hour of bark beetle attack until several weeks after attack, and are more efficient than humans in detecting early infestations.
Sanitation	By far, sanitation is the most frequent and effective direct control method for European spruce beetle. In many countries, infested trees are located by systematic searching on foot and marked for removal by trained technicians fully employed throughout most of the year. Trees recently damaged by windstorms or other disturbances are also often salvaged at this time (Göthlin et al., 2000).
Insecticides	The use of insecticides for management of European spruce beetle is highly restricted in most countries. During sanitation, insecticides may be used to treat infested logs stored along forest roads if they cannot be removed prior to beetle emergence.
Semiochemicals	The use of semiochemicals includes applications of aggregation pheromones and co-attractants (e.g., host volatiles) for survey and detection, mass trapping (trap-out), and trap-tree methods. Repellents (inhibitors) are not widely used (Fettig and Hilszczański, 2015).
Trap trees	Damaged and windthrown trees are often used as trap trees. Trees may be baited with synthetic pheromones and are several times more effective at trapping European spruce beetle than pheromone-baited traps (Raty et al., 1995).
Thinning	Thinning has long been advocated as a preventive measure (Wermelinger, 2004).
Increasing landscape heterogeneity	Increasing heterogeneity (e.g., of age, size, and tree species compositions) and diversity increases resistance of forested landscapes (Wermelinger, 2004).
Reducing the rate of atmospheric warming	Forests are increasingly vulnerable to European spruce beetle as a result of the direct and indirect effects of climate change (Marini et al., 2017).

subspecies near shipping ports of entry in western North America (Myers et al., 2000; Gray, 2017). These ecologically and behaviorally distinct subspecies were at first distinguished by mitochondrial and nuclear DNA sequencing techniques, including microsatellite DNA analysis (Bogdanowicz et al., 1997, 2000), but have been analyzed more recently with increasingly sophisticated approaches and methodologies (Islam et al., 2015; Stewart et al., 2016; Djoumad et al., 2017). With its broader host range (that includes conifers) and the flight capacity of females, Asian gyspy moth is considered a greater threat

Table 8 Methods of control for European spruce beetle during outbreaks in southern Poland, 2007-2010

Method	2007	2008	2009	2010	Total
Trap trees (thousands)	8.4	5.2	4.4	3.7	21.7
Baited-trap trees (thousands of trees, m^3)	17.5	30.9	31.7	16.2	96.3
Marked (infested) trees (thousands)	510	424	272	158	1364
Sanitation (thousands of trees, m^3)	803	798	466	231	2298
Debarked-infested trees (thousands of trees, m^3)	297	254	17	23	591
Pheromone-baited traps (thousands)	11.6	12.2	11.8	10.6	46.2

Source: based on Szabla (2013); adapted from Fettig and Hilszczański (2015).

to the forests of western North America and elsewhere outside of its native range (Peterson et al., 2007). Both subspecies continue to be detected in a network of sex pheromone-baited traps in North America (Tobin and Liebhold, 2011; CFPC, 2010-2017).

7.5.1 Advances in population biology

The long-established populations of European gypsy moth in eastern North America have provided opportunities for research on an invasive hardwood defoliator with periodic outbreak potential (Liebhold et al., 2007; Liebhold and Tobin, 2008; Grayson and Johnson, 2018). The original introduced population has spread steadily from Massachusetts, USA, to include most of the northeastern United States, as well as portions of the midwestern and southeastern United States. The areas of research coupled with this invasion have included local establishment and spread of populations, dynamics, and regulation of populations, impacts of populations, and multiple aspects of management (Table 9).

7.5.2 Management by slowing the rate of population spread

Managing populations of European gypsy moth in North America has involved three approaches: suppression of outbreak populations within the generally infested area; eradication of satellite populations detected at great distance from the generally infested area (typically in the western United States or Canada); and attenuating the spread of populations from the generally infested area. The latter approach, "Slow the Spread" (STS) is a US state-federal cooperative program, whose objective is to retard the spread of European gypsy moth into new areas (Leuschner et al., 1996; Sharov et al., 1998, 2002b; Tobin et al., 2004; McManus, 2007). Several regional integrated pest management programs targeting gypsy moth were the progenitors for STS, which debuted on a small

Table 9 Research advances in the population biology and management of European gypsy moth in North America

Establishment and spread of populations	Fundamental research on the establishment and spread of invasive populations has provided the basis for management of the range expansion of *L. d. dispar* in North America (Sharov et al., 1996b, 1998; Sharov and Liebhold, 1998; Johnson et al., 2006; Liebhold and Tobin, 2006; Liebhold et al., 2007; Tobin et al., 2015; Streifel et al., 2017).
Dynamics and regulation of populations	Low-density populations of *L. d. dispar* appear to be loosely regulated in an inversely density-dependent manner through predation by small mammals; one of the largest sources of mortality in most *L. d. dispar* populations is the fungal pathogen *Entomophaga maimaiga*, which acts largely in a density-independent fashion (Elkinton and Liebhold, 1990; Liebhold, 1992; Liebhold et al., 2000; Tobin and Hajek, 2012; A.M. Liebhold, personal correspondence).
Impact	Partial or total defoliation of oak canopies by *L. d. dispar* often result in tree growth loss, severe physiological stress, and, when exerted over consecutive years or with other sources of stress, tree death. Greatest economic impacts occur in residential areas, where loss of foliage and occasional tree mortality are compounded by the considerable nuisance of large quantities of caterpillars and their urticating hairs and frass (Muzika and Liebhold, 1999, 2000; Morin et al., 2004; Tobin and Liebhold, 2011; Morin and Liebhold, 2016; Liebhold et al., 2017).
Management-Forecasting and surveillance	Data from geographic information systems, forest inventory and analysis, and geospatial methods are used for forecasting defoliation by *L. d. dispar* (Liebhold et al., 1998; Liebhold and Tobin, 2008; Régnière et al., 2009).
Management-Treatments	Treatment efficacy is evaluated by using data from geographic information systems (Liebhold et al., 1996) or egg mass surveys (Liebhold et al., 1991). Populations are controlled through mating disruption with sex pheromone (Leonhardt et al., 1996; Onufrieva et al., 2019). Insecticides (Liebhold and McManus, 1999) and silvicultural treatments (Muzika and Liebhold, 2000; Muzika, 2017) have also been used in management.
Management-Execution and evaluation of the Slow-the-Spread program	An intensive Slow-the-Spread program has been implemented successfully in the United States for nearly 20 years (see text for details (Leonard and Sharov, 1995; Leuschner et al., 1996; Sharov et al., 1998, 2002a,b; Sharov and Liebhold, 1998; Leonard, 2007; Tobin and Blackburn, 2007)).

scale in 1992 (in four US states: Michigan, North Carolina, Virginia, and West Virginia), but was expanded and became operational in 2000 to include the entire expanding front (McManus, 2007). The theoretical underpinnings of the program include a reliance on historical data analysis of the invasive spread of European gypsy moth in North America (Liebhold et al., 2007) and the realization that spread has been facilitated by processes such as stratified diffusion or colony coalescence, whereby accidental movement of life stages by humans ahead of the invading population front initiate isolated population foci that coalesce and expand the front (Sharov and Liebhold, 1998; Liebhold et al., 2007). This process can be disrupted (slowed) by annual aggressive aerial treatments at newly discovered foci, primarily with sex pheromone (mating disruption) targeting adult males, or to a lesser extent with *Btk* targeting larvae (Onufrieva et al., 2019). Detection trapping of male moths (Sharov et al., 1996a, 2002b) is critical for characterizing the new foci for treatment. The modern gypsy moth STS program is administered and executed through a foundation (Gypsy Moth Slow the Spread Foundation, Inc.; McManus, 2007) with technical leadership provided by the USDA Forest Service (http://www.gmsts.org/).

8 Conclusion and future trends

A large body of recent research has quantified the direct response of native and invasive forest insects to recent changing temperatures, in addition to indirect effects due to changes in temperature and precipitation patterns that stress host trees. In many cases, recent climatic changes have been positively correlated with range shifts northward and increased tree mortality within historic ranges. Predictions of future responses are mixed depending on the insect species and type of model used (i.e., correlative or process-based). Although warming temperatures drive model predictions for northward range expansion in most cases, range retractions of southern distributions are predicted for some species, a result of disruptions in cues for seasonality traits such as diapause and developmental rates and thresholds. Adequate predictions of future range expansions and insect-associated tree mortality events will benefit from research that quantifies (1) thermally dependent traits for use in process-based projection models, (2) climate effects on critical community associates and trophic interactions, and (3) adaptive potential of forest insects in a rapidly changing climate.

Describing and harnessing heritable genetic variation in tree resistance to native and invasive insects are becoming recognized as important aspects of future forest management. Recent advances in resistance screening techniques based on non-destructive spectrometric screening tools and modern -omics approaches provide avenues for phenotyping in the field. Research focused on intensive resistance screening can facilitate tree breeding programs to exploit naturally occurring variation in traits that confer resistance to specific native and invasive species.

Survey and detection are critical aspects of native and invasive forest insect management, including early detection of new outbreak foci so that tactics can be applied to maintain or reduce population levels below Allee thresholds for mating success. This is particularly important in slowing the spread of invaders (see below) and native defoliators. A variety of techniques have proven successful including the application of semiochemicals, used in conjunction with traps of varying colors and types, girdled trap trees, and genomic techniques for enhanced identification and monitoring of invasive species. The use of drone aircraft equipped with high-definition sensors and photographic equipment offers a new platform for survey and detection of forest insect-caused tree mortality.

Following detection, integrated biocontrol strategies are currently being evaluated for their direct impact on forest insect populations. With invasive insects in particular, strategies are aimed at disrupting, or slowing, population expansion fronts. For example, a recent large pilot study to reduce spread of the invasive emerald ash borer, SLAM, employs a combination of tactics including girdled ash trees, regulatory control, public outreach, tree removal, and insecticide treatments. For defoliators, aerial spray treatments with multiple products, including sex pheromones for mating disruption and a bacterium *Bacillus thuringiensis* that causes larval mortality, have been shown to suppress specific populations in North America. Release of natural enemies has also proven successful for some native and invasive defoliator species in North America and Europe. Similar tactics are showing promise or are under evaluation for management of native and invasive bark beetle species, including release of predatory beetles (*Rhizophagus grandis*) in Europe and China, and use of the fungus *Beauveria bassiana* for controlling populations of *Dendroctonus* in North America and the European spruce beetle in Europe. Other direct control measures for bark beetles include new formulations and injection methods of insecticides, and development and commercialization of semiochemical formulations and delivery methods to disrupt adult aggregation.

Indirect control measures using silvicultural strategies have also proven efficacious for regulating multiple forest insect species, including non-native invaders. Recent research suggests that thinning stands is effective for increasing resistance to several bark beetle species in North America even during significant drought years. Maintaining high tree age and species diversity, including planting of tree species that are not hosts for particular insect species, has been associated with slower expansion rates of defoliators. Following successful tree breeding programs, planting and maintaining resistant genotypes on the landscape will be a long-term strategy for managing native and invasive forest insect species.

9 Where to look for further information

- A few general references describing climate effects on forest insects, including predictions and management options, are described in Bentz et al. (2010, 2019), Netherer and Schopf (2010), Règniére et al. (2012), Weed et al. (2013), Ayres and Lombardero (2018), and Pureswaran et al. (2018).
- The impact of drought on tree mortality and forest insects is reviewed by Jactel et al. (2012), Allen et al. (2015), and Kolb et al. (2016).

- Garnas (2018) provides a review of insect plasticity and potential response to rapid climate change.
- The importance of understanding host tree resistance in invasive insect-plant interactions is summarized in Villari et al. (2016) and Schowalter et al. (2018), and Telford et al. (2015) and Sniezko and Koch (2017) provide reviews of harnessing natural genetic variation for breeding trees and managing forest insects.
- Aukema et al. (2010, 2011), Liebhold et al. (2017), and Poland et al. (2019) provide overviews of invasive forest insect biology, ecology, impact, and management.
- Fettig and Hilszczański (2015) provide an overview of management strategies for conifer bark beetles.
- Seybold et al. (2018) provide a synthesis of the concepts involved in semiochemical-based management of bark beetles with case studies of western North American species.

10 References

Abdullah, H., Skidmore, A. K., Darvishzadeh, R. and Heurich, M. 2019. Sentinel-2 accurately maps green-attack stage of European spruce bark beetle (*Ips typographus* L.) compared with Landsat-8. *Remote Sens. Ecol. Conserv.* 5(1), 87-106. doi:10.1002/rse2.93.

Ackermann, J., Adler, P., Hoffmann, K., Hurling, R., John, R., Otto, L., Sagischewski, H., Seitz, R., Straub, C. and Stürtz, M. 2018. Früherkennung von Buchdruckerbefall durch Drohnen (Early detection of *Ips typographus* infestation by drones). *AFZ-DerWald* 19, 50-3.

Addison, A., Powell, J. A., Bentz, B. J. and Six, D. L. 2015. Integrating models to investigate critical phenological overlaps in complex ecological interactions: the mountain pine beetle-fungus symbiosis. *J. Theor. Biol.* 368, 55-66. doi:10.1016/j.jtbi.2014.12.011.

Aflitto, N. C. and Hofstetter, R. W. 2014. Use of acoustics to deter bark beetles from entering tree material. *Pest Manag. Sci.* 70(12), 1808-14. doi:10.1002/ps.3720.

Alfaro, R. I., vanAkker, L., Jaquish, B. and King, J. 2004. Weevil resistance of progeny derived from putatively resistant and susceptible interior spruce parents. *For. Ecol. Manag.* 202(1-3), 369-77. doi:10.1016/j.foreco.2004.08.001.

Alfaro, R. I., King, J. N. and vanAkker, L. 2013. Delivering Sitka spruce with resistance against white pine weevil in British Columbia, Canada. *For. Chron.* 89(2), 235-45. doi:10.5558/tfc2013-042.

Allen, C. D., Breshears, D. D. and McDowell, N. G. 2015. On underestimation of global vulnerability to tree mortality and forest die-off from hotter drought in the Anthropocene. *Ecosphere* 6, 1-55.

Anderegg, W. R. L., Kane, J. M. and Anderegg, L. D. L. 2013. Consequences of widespread tree mortality triggered by drought and temperature stress. *Nat. Clim. Chang.* 3(1), 30-6. doi:10.1038/nclimate1635.

Arora, V. K., Peng, Y., Kurz, W. A., Fyfe, J. C., Hawkins, B. and Werner, A. T. 2016. Potential near-future carbon uptake overcomes losses from a large insect outbreak in British Columbia, Canada. *Geophys. Res. Lett.* 43(6), 2590-8. doi:10.1002/2015GL067532.

Aukema, J. E., McCullough, D. G., Von Holle, B., Liebhold, A. M., Britton, K. and Frankel, S. J. 2010. Historical accumulation of nonindigenous forest pests in the continental United States. *BioScience* 60(11), 886–97. doi:10.1525/bio.2010.60.11.5.

Aukema, J. E., Leung, B., Kovacs, K., Chivers, C., Britton, K. O., Englin, J., Frankel, S. J., Haight, R. G., Holmes, T. P., Liebhold, A. M., McCullough, D. G. and Von Holle, B. 2011. Economic impacts of non-native forest insects in the continental United States. *PLoS ONE* 6(9), e24587. doi:10.1371/journal.pone.0024587.

Ayres, M. P. and Lombardero, M. J. 2018. Forest pests and their management in the Anthropocene. *Can. J. For. Res.* 48(3), 292–301. doi:10.1139/cjfr-2017-0033.

Baier, P., Pennerstorfer, J. and Schopf, A. 2007. PHENIPS–a comprehensive phenology model of *Ips typographus* (L.)(Col., Scolytinae) as a tool for hazard rating of bark beetle infestation. *For. Ecol. Manag.* 249(3), 171–86. doi:10.1016/j.foreco.2007.05.020.

Bale, J. S., Masters, G. J., Hodkinson, I. D., Awmack, C., Bezemer, T. M., Brown, V. K., Butterfield, J., Buse, A., Coulson, J. C., Farrar, J., Good, J. E. G., Harrington, R., Hartley, S., Jones, T. H., Lindroth, R. L., Press, M. C., Symrnioudis, I., Watt, A. D. and Whittaker, J. B. 2002. Herbivory in global climate change research: direct effects of rising temperature on insect herbivores. *Glob. Chang. Biol.* 8(1), 1-16. doi:10.1046/j.1365-2486.2002.00451.x.

Baranchikov, Y., Akulov, E. and Astapenko, S. 2010. Bark beetle *Polygraphus proximus*: a new aggressive far eastern invader on *Abies* species in Siberia and European Russia. USDA Research Forum on Invasive Species GTR-NRS-P-75. Available at: http://www.treesearch.fs.fed.us/pubs/37559.

Barham, E. 2016. The unique role of sentinel trees, botanic gardens and arboreta in safeguarding global plant health. *Plant Biosyst.* 150(3), 377–80. doi:10.1080/11263504.2016.1179231.

Barredo, J. I., Strona, G., de Rigo, D., Caudullo, G., Stancanelli, G. and San-Miguel-Ayanz, J. 2015. Assessing the potential distribution of insect pests: case studies on large pine weevil (*Hylobius abietis* L.) and horse-chestnut leaf miner (*Cameraria ohridella*) under present and future climate conditions in European forests. *EPPO Bull.* 45(2), 273-81. doi:10.1111/epp.12208.

Barta, M., Kautmanová, I., Čičková, H., Ferenčík, J., Florián, S., Novotný, J. and Kozánek, M. 2018. The potential of *Beauveria bassiana* inoculum formulated into a polymeric matrix for a microbial control of spruce bark beetle. *Bio. Sci. Tech.* 28(7), 718–35. doi:10.1080/09583157.2018.1487027.

Battisti, A., Stastny, M., Netherer, S., Robinet, C., Schopf, A., Roques, A. and Larsson, S. 2005. Expansion of geographic range in the pine processionary moth caused by increased winter temperatures. *Ecol. Appl.* 15(6), 2084–96. doi:10.1890/04-1903.

Battisti, A., Benvegnù, I., Colombari, F. and Haack, R. A. 2014. Invasion by the chestnut gall wasp in Italy causes significant yield loss in *Castanea sativa* nut production. *Ag. For. Entomol.* 16(1), 75–9. doi:10.1111/afe.12036.

Bauer, L. S., Duan, J. J., Gould, J. R. and Van Driesche, R. G. 2015. Progress in the classical biological control of *Agrilus planipennis* Fairmaire (Coleoptera: Buprestidae) in North America. *Can. Entomol.* 147(3), 300–17. doi:10.4039/tce.2015.18.

Bentz, B. J. and Hansen, E. M. 2018. Evidence for a prepupal diapause in the mountain pine beetle (*Dendroctonus ponderosae*). *Environ. Entomol.* 47(1), 175–83. doi:10.1093/ee/nvx192.

Bentz, B. J. and Jönsson, A. M. 2015. Modeling bark beetle responses to climate change. In: Vega, F. and Hofstetter, R. (Eds), *Bark Beetles, Biology and Ecology of Native and*

Invasive Species. Elsevier, London and Oxford, UK, San Diego, CA, Waltham, MA, pp. 533–53. Chapter 13.

Bentz, B. J., Régnière, J., Fettig, C. J., Hansen, E. M., Hayes, J. L., Hicke, J. A., Kelsey, R. G., Negrón, J. F. and Seybold, S. J. 2010. Climate change and bark beetles of the western United States and Canada: direct and indirect effects. *BioScience* 60(8), 602–13. doi:10.1525/bio.2010.60.8.6.

Bentz, B. J., Bracewell, R. R., Mock, K. E. and Pfrender, M. E. 2011. Genetic architecture and phenotypic plasticity of thermally-regulated traits in an eruptive species, *Dendroctonus ponderosae*. *Evol. Ecol.* 25(6), 1269–88. doi:10.1007/s10682-011-9474-x.

Bentz, B. J., Duncan, J. P. and Powell, J. A. 2016. Elevational shifts in thermal suitability for mountain pine beetle population growth in a changing climate. *Forestry* 89(3), 271–83. doi:10.1093/forestry/cpv054.

Bentz, B. J., Hood, S. M., Hansen, E. M., Vandygriff, J. C. and Mock, K. E. 2017. Defense traits in the long-lived Great Basin bristlecone pine and resistance to the native herbivore mountain pine beetle. *New Phytol.* 213(2), 611–24. doi:10.1111/nph.14191.

Bentz, B. J., Jönsson, A. M., Schroeder, M., Weed, A., Wilcke, R. A. I. and Larsson, K. 2019. *Ips typographus* and *Dendroctonus ponderosae* models project thermal suitability for intra-and inter-continental establishment in a changing climate. *Front. For. Glob. Chang.* doi:10.3389/ffgc.2019.00001.

Berg, E. E., Henry, J. D., Fastie, C. L., De Volder, A. D. and Matsuoka, S. M. 2006. Spruce beetle outbreaks on the Kenai Peninsula, Alaska, and Kluane National Park and Reserve, Yukon Territory: relationship to summer temperatures and regional differences in disturbance regimes. *For. Ecol. Manag.* 227(3), 219–32. doi:10.1016/j.foreco.2006.02.038.

Berner, L. T., Law, B. E., Meddens, A. J. H. and Hicke, J. A. 2017. Tree mortality from fires, bark beetles, and timber harvest during a hot and dry decade in the western United States (2003–2012). *Environ. Res. Lett.* 12(6). doi:10.1088/1748-9326/aa6f94.

Berryman, A. A., Amman, G. D. and Stark, R. W. 1978. *Theory and Practice of Mountain Pine Beetle Management in Lodgepole Pine Forests*. Washington State University, Pullman, WA.

Bilodeau, P., Roe, A. D., Bilodeau, G., Blackburn, G. S., Cui, M., Cusson, M., Doucet, D., Griess, V. C., Lafond, V. M. A., Nilausen, C., Paradis, G., Porth, I., Prunier, J., Srivastava, V., Stewart, D., Torson, A. S., Tremblay, E., Uzunovic, A., Yemshanov, D. and Hamelin, R. C. 2019. Biosurveillance of forest insects: part II–adoption of genomic tools by end user communities and barriers to integration. *J. Pest. Sci.* 92(1), 71–82. doi:10.1007/s10340-018-1001-1.

Björkman, C., Bylund, H., Nilsson, U., Nordlander, G. and Schroeder, M. 2015. Effects of new forest management on insect damage risk in a changing climate. In: Björkman, C. and Niemelä, P. (Eds), *Climate Change and Insect Pests*. CABI Publishing, Wallingford, UK, pp. 248–66. Chapter 14.

Bogdanowicz, S. M., Mastro, V. C., Prasher, D. C. and Harrison, R. G. 1997. Microsatellite DNA variation among Asian and North American gypsy moths (Lepidoptera: Lymantriidae). *Ann. Entomol. Soc. Am.* 90(6), 768–75. doi:10.1093/aesa/90.6.768.

Bogdanowicz, S. M., Schaefer, P. W. and Harrison, R. G. 2000. Mitochondrial DNA variation among worldwide populations of gypsy moths, *Lymantria dispar*. *Mol. Phylogenet. Evol.* 15(3), 487–95. doi:10.1006/mpev.1999.0744.

Bozsik, G. and Szőcs, G. 2017. Phenology, behavior and infestation levels of the invasive small cypress bark beetle, *Phloeosinus aubei*, on some cultivars of *Thuja* and *Juniperus* spp., in Hungary. *Phytoparasitica* 45(2), 201–10. doi:10.1007/s12600-017-0585-y.

Brasier, C. M. and Buck, K. W. 2001. Rapid evolutionary changes in a globally invading fungal pathogen (Dutch elm disease). *Biol. Invas.* 3(3), 223–33. doi:10.1023/A:1015248819864.

Brecka, A. F. J., Shahi, C. and Chen, H. Y. H. 2018. Climate change impacts on boreal forest timber supply. *For. Policy Econ.* 92, 11–21. doi:10.1016/j.forpol.2018.03.010.

Brockerhoff, E. G. and Liebhold, A. M. 2017. Ecology of forest insect invasions. *Biol. Invas.* 19(11), 3141–59. doi:10.1007/s10530-017-1514-1.

Buotte, P. C., Hicke, J. A., Preisler, H. K., Abatzoglou, J. T., Raffa, K. F. and Logan, J. A. 2016. Climate influences on whitebark pine mortality from mountain pine beetle in the Greater Yellowstone Ecosystem. *Ecol. Appl.* 26(8), 2505–22. doi:10.1002/eap.1396.

Butin, E., Porter, A. H. and Elkinton, J. 2005. Adaptation during biological invasions and the case of *Adelges tsugae*. *Evol. Ecol. Res.* 7, 887–900.

California Forest Pest Council (CFPC). 2010–2017. California forest pest conditions 2010-2017. California Forest Pest Council, Sacramento, CA. Available at: https://www.fs.usda.gov/detail/r5/forest-grasslandhealth/?cid=fsbdev3_046704 (accessed on 11 February 2019).

Campbell, E. M., MacLean, D. A. and Bergeron, Y. 2008. The severity of budworm-caused growth reductions in balsam fir/spruce stands varies with the hardwood content of surrounding forest landscapes. *For. Sci.* 54, 195–205.

Candau, J.-N. and Fleming, R. A. 2005. Landscape-scale spatial distribution of spruce budworm defoliation in relation to bioclimatic conditions. *Can. J. For. Res.* 35(9), 2218–32. doi:10.1139/x05-078.

Candau, J.-N. and Fleming, R. A. 2011. Forecasting the response of spruce budworm defoliation to climate change in Ontario. *Can. J. For. Res.* 41(10), 1948–60. doi:10.1139/x11-134.

Carnegie, A. J., Matsuki, M., Haugen, D. A., Hurley, B. P., Ahumada, R., Klasmer, P., Sun, J. and Iede, E. T. 2006. Predicting the potential distribution of *Sirex noctilio* (Hymenoptera: Siricidae), a significant exotic pest of *Pinus* plantations. *Ann. For. Sci.* 63(2), 119–28. doi:10.1051/forest:2005104.

Carroll, A., Taylor, S., Régnière, J. and Safranyik, L. 2004. Effects of climate and climate change on the mountain pine beetle. In: *Proceedings of the Mountain Pine Beetle Symposium: Challenges and Solutions*. Information Rep. BC-X-399. Canadian Forest Service, Pacific Forestry Centre, pp. 223–32.

Chen, I. C., Hill, J. K., Ohlemüller, R., Roy, D. B. and Thomas, C. D. 2011. Rapid range shifts of species associated with high levels of climate warming. *Science* 333(6045), 1024–6. doi:10.1126/science.1206432.

Chen, Y., Vasseur, L. and You, M. 2017. Potential distribution of the invasive loblolly pine mealybug, *Oracella acuta* (Hemiptera: Pseudococcidae), in Asia under future climate change scenarios. *Clim. Chang.* 141(4), 719–32. doi:10.1007/s10584-017-1917-0.

Chou, H. H. and Keasling, J. D. 2012. Synthetic pathway for production of five-carbon alcohols from isopentenyl diphosphate. *Appl. Environ. Microbiol.* 78(22), 7849–55. doi:10.1128/AEM.01175-12.

Ciesla, W. 2011. *Forest Entomology: a Global Perspective*. John Wiley & Sons, West Sussex, UK.

Coleman, T. W. and Seybold, S. J. 2008. Previously unrecorded damage to oak, *Quercus* spp., in southern California by the goldspotted oak borer, *Agrilus coxalis* Waterhouse (Coleoptera: Buprestidae). *Pan-Pac. Entomol.* 84(4), 288–300. doi:10.3956/2008-18.1.

Coleman, T. W., Graves, A. D., Hoddle, M. S., Heath, Z., Flint, M. L., Chen, Y. and Seybold, S. J. 2012. Forest stand composition and impacts associated with *Agrilus auroguttatus* Schaeffer (Coleoptera: Buprestidae) and *Agrilus coxalis* Waterhouse in oak woodlands. *For. Ecol. Manag.* 276 (2012), 104–17. http://dx.doi.org/10.1016/j.foreco.2012.03.011

Coleman, T. W., Poloni, A. L., Chen, Y., Thu, P.-Q., Li, Q., Sun, J.-H., Rabaglia, R. J., Man, G. and Seybold, S. J. 2019. Hardwood injury and mortality associated with two shot hole borers, *Euwallacea* spp., in the invaded region of southern California, USA and the native region of Southeast Asia. *Ann. For. Sci.* (https://doi.org/10.1007/s13595-019-0847-6).

Conrad, A. O. and Bonello, P. 2016. Application of infrared and Raman spectroscopy for the identification of disease resistant trees. *Front. Plant Sci.* 6, 1152. doi:10.3389/fpls.2015.01152.

Conrad, A. O., Rodriguez-Saona, L. E., McPherson, B. A., Wood, D. L. and Bonello, P. 2014. Identification of *Quercus agrifolia* (coast live oak) resistant to the invasive pathogen *Phytophthora ramorum* in native stands using Fourier-transform infrared (FT-IR) spectroscopy. *Front. Plant Sci.* 5, 521. doi:10.3389/fpls.2014.00521.

Cooke, B. J. and Carroll, A. L. 2017. Predicting the risk of mountain pine beetle spread to eastern pine forests: considering uncertainty in uncertain times. *For. Ecol. Manag.* 396, 11–25. doi:10.1016/j.foreco.2017.04.008.

Cooke, B. J. and Roland, J. 2018. Early 20th century climate-driven shift in the dynamics of forest tent caterpillar outbreaks. *Am. J. Clim. Change* 7, 253.

Cooke, B. J., Nealis, V. G. and Régnière, J. 2007. Insect defoliators as periodic disturbances in northern forest ecosystems. In: Johnson, E. and Miyanishi, K. (Eds), *Plant Disturbance Ecology: the Process and the Response*. Elsevier Academic Press, Burlington, MA, pp. 487–525.

Crook, D. J., Khrimian, A., Francese, J. A., Fraser, I., Poland, T. M., Sawyer, A. J. and Mastro, V. C. 2008. Development of a host-based semiochemical lure for trapping emerald ash borer, *Agrilus planipennis* (Coleoptera: Buprestidae). *Environ. Entomol.* 37(2), 356–65. doi:10.1603/0046-225x(2008)37[356:doahsl]2.0.co;2.

Crowther, T. W., Glick, H. B., Covey, K. R., Bettigole, C., Maynard, D. S., Thomas, S. M., Smith, J. R., Hintler, G., Duguid, M. C., Amatulli, G., Tuanmu, M. N., Jetz, W., Salas, C., Stam, C., Piotto, D., Tavani, R., Green, S., Bruce, G., Williams, S. J., Wiser, S. K., Huber, M. O., Hengeveld, G. M., Nabuurs, G. J., Tikhonova, E., Borchardt, P., Li, C. F., Powrie, L. W., Fischer, M., Hemp, A., Homeier, J., Cho, P., Vibrans, A. C., Umunay, P. M., Piao, S. L., Rowe, C. W., Ashton, M. S., Crane, P. R. and Bradford, M. A. 2015. Mapping tree density at a global scale. *Nature* 525(7568), 201–5. doi:10.1038/nature14967.

Cruz, M. G. and Alexander, M. E. 2010. Assessing crown fire potential in coniferous forests of western North America: A critique of current approaches and recent simulation studies. *Int. J. Wildland Fire* 19(4), 377–98. doi:10.1071/WF08132.

Cuddington, K., Sobek-Swant, S., Crosthwaite, J. C., Lyons, D. B. and Sinclair, B. J. 2018. Probability of emerald ash borer impact for Canadian cities and North America: a mechanistic model. *Biol. Invas.* 20(9), 2661–77. doi:10.1007/s10530-018-1725-0.

Daoust, S. P., Mader, B. J., Bauce, E., Despland, E., Dussutour, A. and Albert, P. J. 2010. Influence of epicuticular-wax composition on the feeding pattern of a phytophagous insect: implications for host resistance. *Can. Entomol.* 142(3), 261–70. doi:10.4039/n09-064.

Dara, S. K., Barringer, L. and Arthurs, S. P. 2015. *Lycorma delicatula* (Hemiptera: Fulgoridae): a new invasive pest in the United States. *J. Integ. Pest Manag.* 6(1), 20. doi:10.1093/jipm/pmv021.

Davídková, M. and Doležal, P. 2017. Sister broods in the spruce bark beetle, *Ips typographus* (L.). *For. Ecol. Manag.* 405, 13–21. doi:10.1016/j.foreco.2017.08.040.

Davidson, C. B., Gottschalk, K. W. and Johnson, J. E. 2001. European gypsy moth (*Lymantria dispar* L.) outbreaks: a review of the literature. Gen. Tech. Rep. NE-278. United States Department of Agriculture, Forest Service, Northeastern Research Station, Newtown Square, PA, 15pp.

Davis, T. S., Mann, A. J., Malesky, D., Jankowski, E. and Bradley, C. 2018. Laboratory and field evaluation of the entomopathogenic fungus *Beauveria bassiana* (Deuteromycotina: Hyphomycetes) for population management of spruce beetle, *Dendroctonus rufipennis* (Coleoptera: Scolytinae), in felled trees and factors limiting pathogen success. *Environ. Entomol.* 47(3), 594–602. doi:10.1093/ee/nvy036.

de Groot, P., Grant, G. G., Poland, T. M., Scharbach, R., Buchan, L., Nott, R. W., Macdonald, L. and Pitt, D. 2008. Electrophysiological response and attraction of emerald ash borer to green leaf volatiles (GLVs) emitted by host foliage. *J. Chem. Ecol.* 34(9), 1170–9. doi:10.1007/s10886-008-9514-3.

Delvas, N., Bauce, É., Labbé, C., Ollevier, T. and Bélanger, R. 2011. Phenolic compounds that confer resistance to spruce budworm. *Entomol. Exp. Appl.* 141(1), 35–44. doi:10.1111/j.1570-7458.2011.01161.x.

DeRose, R. J., Bentz, B. J., Long, J. N. and Shaw, J. D. 2013. Effect of increasing temperatures on the distribution of spruce beetle in Engelmann spruce forests of the Interior West, USA. *For. Ecol. Manag.* 308, 198–206. doi:10.1016/j.foreco.2013.07.061.

Djoumad, A., Nisole, A., Zahiri, R., Freschi, L., Picq, S., Gundersen-Rindal, D. E., Sparks, M. E., Dewar, K., Stewart, D., Maaroufi, H., Levesque, R. C., Hamelin, R. C. and Cusson, M. 2017. Comparative analysis of mitochondrial genomes of geographic variants of the gypsy moth, *Lymantria dispar*, reveals a previously undescribed genotypic entity. *Sci. Rep.* 7(1), 14245. doi:10.1038/s41598-017-14530-6.

Dodds, K. J. and Orwig, D. A. 2011. An invasive urban forest pest invades natural environments–Asian longhorned beetle in northeastern US hardwood forests. *Can. J. For. Res.* 41(9), 1729–42. doi:10.1139/x11-097.

Dodds, K. J., Cooke, R. R. and Hanavan, R. P. 2014. The effects of silvicultural treatment on *Sirex noctilio* attacks and tree health in northeastern United States. *Forests* 5(11), 2810–24. doi:10.3390/f5112810.

Dodds, K. J., Aoki, C. F., Arango-Velez, A., Cancelliere, J., D'Amato, A. W., DiGirolomo, M. F. and Rabaglia, R. J. 2018. Expansion of southern pine beetle into northeastern forests: management and impact of a primary bark beetle in a new region. *J. For.* 116(2), 178–91. doi:10.1093/jofore/fvx009.

Duan, J. J., Abell, K. J., Bauer, L. S., Gould, J. and Van Driesche, R. 2014. Natural enemies implicated in the regulation of an invasive pest: a life table analysis of the population dynamics of the emerald ash borer. *Ag. For. Entomol.* 16(4), 406–16. doi:10.1111/afe.12070.

Duan, J. J., Bauer, L. S., van Driesche, R. and Gould, J. 2018. Progress and challenges of protecting North American ash trees from the emerald ash borer using biological control. *Forests* 9(3), 142. doi:10.3390/f9030142.

Dugan, A. J., Birdsey, R., Mascorro, V. S., Magnan, M., Smyth, C. E., Olguin, M. and Kurz, W. A. 2018. A systems approach to assess climate change mitigation options in landscapes of the United States forest sector. *Carbon Balance Manag.* 13(1), 13. doi:10.1186/s13021-018-0100-x.

Dukes, J. S., Pontius, J., Orwig, D., Garnas, J. R., Rodgers, V. L., Brazee, N., Cooke, B., Theoharides, K. A., Stange, E. E., Harrington, R., Ehrenfeld, J., Gurevitch, J., Lerdau, M., Stinson, K., Wick, R. and Ayres, M. 2009. Responses of insect pests, pathogens, and invasive plant species to climate change in the forests of northeastern North America: what can we predict?. *Can. J. For. Res.* 39(2), 231–48. doi:10.1139/X08-171.

EAB Info (Emerald Ash Borer Information). 2019. Emerald Ash Borer Information Network. Available at: www.emeraldashborer.info/.

Eidson, E. L., Mock, K. E. and Bentz, B. J. 2017. Mountain pine beetle host selection behavior confirms high resistance in Great Basin bristlecone pine. *For. Ecol. Manag.* 402, 12–20. doi:10.1016/j.foreco.2017.06.034.

Eidson, E. L., Mock, K. E. and Bentz, B. J. 2018. Low offspring survival in mountain pine beetle infesting the resistant Great Basin bristlecone pine supports the preference-performance hypothesis. *PLoS ONE* 13(5), e0196732. doi:10.1371/journal.pone.0196732.

Elkinton, J. S. and Liebhold, A. M. 1990. Population dynamics of gypsy moth in North America. *Annu. Rev. Entomol.* 35(1), 571–96. doi:10.1146/annurev.en.35.010190.003035.

EPPO. 2016. PM 7/129 (1) DNA barcoding as an identification tool for a number of regulated pests. *EPPO Bull.* 46(3), 501–37. doi:10.1111/epp.12344.

Erbilgin, N. 2019. Phytochemicals as mediators for host range expansion of a native invasive forest insect herbivore. *New Phytol.* 221(3), 1268–78. doi:10.1111/nph.15467.

Eskalen, A., Stouthamer, R., Lynch, S. C., Rugman-Jones, P. F., Twizeyimana, M., Gonzalez, A. and Thibault, T. 2013. Host range of Fusarium dieback and its ambrosia beetle (Coleoptera: Scolytinae) vector in Southern California. *Plant Dis.* 97(7), 938–51. doi:10.1094/PDIS-11-12-1026-RE.

Esper, J., Büntgen, U., Frank, D. C., Nievergelt, D. and Liebhold, A. 2006. 1200 years of regular outbreaks in alpine insects. *Proc. R. Soc. B* 274, 671–9.

Eveleigh, E. S., McCann, K. S., McCarthy, P. C., Pollock, S. J., Lucarotti, C. J., Morin, B., McDougall, G. A., Strongman, D. B., Huber, J. T., Umbanhowar, J. and Faria, L. D. 2007. Fluctuations in density of an outbreak species drive diversity cascades in food webs. *Proc. Natl. Acad. Sci. U. S. A.* 104(43), 16976–81. doi:10.1073/pnas.0704301104.

Eveleigh, E. S., Lucarotti, C. J., McCarthy, P. C. and Morin, B. 2012. Prevalence, transmission and mortality associated with *Nosema fumiferanae* infections in field populations of spruce budworm *Choristoneura fumiferana*. *Ag. For. Entomol.* 14(4), 389–98. doi:10.1111/j.1461-9563.2012.00580.x.

Evenden, M. L. and Silk, P. J. 2016. The influence of Canadian research on semiochemical-based management of forest insect pests in Canada. *Can. Entomol.* 148(S1), S170–209. doi:10.4039/tce.2015.17.

Eyles, A., Robinson, A. P., Smith, D., Carnegie, A., Smith, I., Stone, C. and Mohammed, C. 2011. Quantifying stem growth loss at the tree-level in a *Pinus radiata* plantation to repeated attack by the aphid, *Essigella californica*. *For. Ecol. Manag.* 261(1), 120–7. doi:10.1016/j.foreco.2010.09.039.

Fält-Nardmann, J. J. J., Klemola, T., Roth, M., Ruohomäki, K. and Saikkonen, K. 2016. Northern geometrid forest pests (Lepidoptera: Geometridae) hatch at lower temperatures than their southern conspecifics: implications of climate change. *Eur. J. Entomol.* 113, 337–43. doi:10.14411/eje.2016.043.

Fält-Nardmann, J. J. J., Klemola, T., Ruohomäki, K., Niemelä, P., Roth, M. and Saikkonen, K. 2017. Local adaptations and phenotypic plasticity may render gypsy moth and nun moth future pests in northern European boreal forests. *Can. J. For. Res.* 48(3), 265–76. doi:10.1139/cjfr-2016-0481.

Fält-Nardmann, J. J. J., Ruohomäki, K., Tikkanen, O.-P. and Neuvonen, S. 2018a. Cold hardiness of *Lymantria monacha* and *L. dispar* (Lepidoptera: Erebidae) eggs to extreme winter temperatures: implications for predicting climate change impacts. *Ecol. Entomol.* 43(4), 422–30. doi:10.1111/een.12515.

Fält-Nardmann, J. J. J., Tikkanen, O.-P., Ruohomäki, K., Otto, L.-F., Leinonen, R., Pöyry, J., Saikkonen, K. and Neuvonen, S. 2018b. The recent northward expansion of *Lymantria monacha* in relation to realised changes in temperatures of different seasons. *For. Ecol. Manag.* 427, 96–105. doi:10.1016/j.foreco.2018.05.053.

FAO. 2018. *The State of the World's Forests 2018- Forest Pathways to Sustainable Development*. Licence: CC BY-NC-SA 3.0 IGO. FAO, Rome.

Feng, Y., Liang, J., Lu, Q. and Zhang, X. 2009. Potential suitability analysis of *Hemiberlesia pitysophila* Takagi in China. *For. Res. Beijing* 22, 563–7.

Fettig, C. J. and Hilszczański, J. 2015. Management strategies for bark beetles in conifer forests. In: Vega, F. E. and Hofstetter, R. W. (Eds), *Bark Beetles: Biology and Ecology of Native and Invasive Species*. Springer, London, pp. 555–84. Chapter 14.

Fettig, C. J. and McKelvey, S. R. 2014. Resiliency of an interior ponderosa pine forest to bark beetle infestations following fuel-reduction and forest-restoration treatments. *Forests* 5(1), 153–76. doi:10.3390/f5010153.

Fettig, C. J., Allen, K. K., Borys, R. R., Christopherson, J., Dabney, C. P., Eager, T. J., Gibson, K. E., Hebertson, E. G., Long, D. F., Munson, A. S., Shea, P. J., Smith, S. L. and Haverty, M. I. 2006a. Effectiveness of bifenthrin (Onyx™) and carbaryl (Sevin® SL) for protecting individual, high-value trees from bark beetle attack (Coleoptera: Curculionidae: Scolytinae) in the western United States. *J. Econ. Entomol.* 99, 1691–98.

Fettig, C. J., DeGomez, T. E., Gibson, K. E., Dabney, C. P. and Borys, R. R. 2006b. Effectiveness of permethrin plus-C (Masterline®) and carbaryl (Sevin SL®) for protecting individual, high-value pines from bark beetle attack. *Arbor. Urban For.* 32, 247–52.

Fettig, C. J., Klepzig, K. D., Billings, R. F., Munson, A. S., Nebeker, T. E., Negrón, J. F. and Nowak, J. T. 2007. The effectiveness of vegetation management practices for prevention and control of bark beetle infestations in coniferous forests of the western and southern United States. *For. Ecol. Manag.* 238(1–3), 24–53. doi:10.1016/j.foreco.2006.10.011.

Fettig, C. J., McKelvey, S. R., Dabney, C. P., Huber, D. P. W., Lait, C. G., Fowler, D. L. and Borden, J. H. 2012a. Efficacy of "verbenone Plus" for protecting ponderosa pine trees and stands from *Dendroctonus brevicomis* (Coleoptera: Curculionidae) attack in British Columbia and California. *J. Econ. Entomol.* 105(5), 1668–80. doi:10.1603/ec12184.

Fettig, C. J., Bulaon, B. M., Dabney, C. P., Hayes, C. J. and McKelvey, S. R. 2012b. Verbenone Plus reduces levels of tree mortality attributed to mountain pine beetle infestations in whitebark pine, a tree species of concern. *J. Biofert. Biopest.* 3, 1–5.

Fettig, C. J., Reid, M. L., Bentz, B. J., Sevanto, S., Spittlehouse, D. L. and Wang, T. 2013a. Changing climates, changing forests: a western North American perspective. *J. For.* 111(3), 214–28. doi:10.5849/jof.12-085.

Fettig, C. J., Grosman, D. M. and Munson, A. S. 2013b. Advances in insecticide tools and tactics for protecting conifers from bark beetle attack in the western United States. In: Trdan, S. (Ed.), *Insecticides - Development of Safer and More Effective Technologies*. InTech, Croatia, pp. 472–92. Chapter 17.

Fettig, C. J., Munson, A. S., Grosman, D. M. and Bush, P. B. 2014a. Evaluations of emamectin benzoate and propiconazole for protecting individual *Pinus contorta* from mortality attributed to colonization by *Dendroctonus ponderosae* and associated fungi. *Pest Manag. Sci.* 70(5), 771–78. doi:10.1002/ps.3612.

Fettig, C. J., Gibson, K. E., Munson, A. S. and Negrón, J. F. 2014b. A comment on "Management for mountain pine beetle outbreak suppression: does relevant science support current policy?". *Forests* 5(4), 822–26. doi:10.3390/f5040822.

Fettig, C. J., Gibson, K. E., Munson, A. S. and Negrón, J. F. 2014c. Cultural practices for prevention and mitigation of mountain pine beetle infestations. *For. Sci.* 60(3), 450–63. doi:10.5849/forsci.13-032.

Fettig, C. J., Munson, A. S. and Gibson, K. E. 2015a. Alternative timing of carbaryl treatments for protecting lodgepole pine from mortality attributed to mountain pine beetle. *Crop Prot.* 69, 56–9. doi:10.1016/j.cropro.2014.12.011.

Fettig, C. J., Munson, A. S., Reinke, M. and Mafra-Neto, A. 2015b. A novel semiochemical tool for protecting lodgepole pine from mortality attributed to mountain pine beetle (Coleoptera: Curculionidae). *J. Econ. Entomol.* 108, 173–82.

Fettig, C. J., Blackford, D. C., Grosman, D. M. and Munson, A. S. 2017. Injections of emamectin benzoate protect Engelmann spruce from mortality attributed to spruce beetle for two years. *J. Entomol. Sci.* 52(2), 193–96. doi:10.18474/JES16-34.1.

Fettig, C. J., Lowrey, L. L., Blackford, D. C., McMillin, J. D., Munson, A. S. and Mortenson, L. A. 2018. Efficacy of spring and fall treatments of carbaryl for protecting ponderosa pine from mortality attributed to mountain pine beetle (Coleoptera: Curculionidae). *J. Econ. Entomol.* 111(6), 2979–82. doi:10.1093/jee/toy259.

Fettig, C. J., Mortenson, L. A., Bulaon, B. M. and Foulk, P. B. 2019. Tree mortality following drought in the central and southern Sierra Nevada, California, U.S. *For. Ecol. Manag.* 432, 164–78. doi:10.1016/j.foreco.2018.09.006.

Fitzpatrick, M. C., Preisser, E. L., Porter, A., Elkinton, J. and Ellison, A. M. 2012. Modeling range dynamics in heterogeneous landscapes: invasion of the hemlock woolly adelgid in eastern North America. *Ecol. App.* 22(2), 472–86. doi:10.1890/11-0009.1.

Flower, A., Gavin, D., Heyerdahl, E., Parsons, R. and Cohn, G. 2014. Drought-triggered western spruce budworm outbreaks in the interior Pacific Northwest: a multi-century dendrochronological record. *For. Ecol. Manag.* 324, 16–27.

Flower, C. E., Knight, K. S. and Gonzalez-Meler, M. A. 2013. Impacts of the emerald ash borer (*Agrilus planipennis* Fairmaire) induced ash (*Fraxinus* spp.) mortality on forest carbon cycling and successional dynamics in the eastern United States. *Biol. Invas.* 15(4), 931–44. doi:10.1007/s10530-012-0341-7.

Formby, J. P., Rodgers, J. C., Koch, F. H., Krishnan, N., Duerr, D. A. and Riggins, J. J. 2018. Cold tolerance and invasive potential of the redbay ambrosia beetle (*Xyleborus glabratus*) in the eastern United States. *Biol. Invas.* 20(4), 995–1007. doi:10.1007/s10530-017-1606-y.

Forrest, J. R. 2016. Complex responses of insect phenology to climate change. *Curr. Opin. Insect Sci.* 17, 49–54. doi:10.1016/j.cois.2016.07.002.

Fox, D. 2007. Ecology. Back to the no-analog future? *Science* 316(5826), 823–5. doi:10.1126/science.316.5826.823.

Fraedrich, S. W., Harrington, T. C., Rabaglia, R. J., Ulyshen, M. D., Mayfield III, A. E., Hanula, J. L., Eickwort, J. M. and Miller, D. R. 2008. A fungal symbiont of the redbay ambrosia beetle causes a lethal wilt in redbay and other Lauraceae in the southeastern United States. *Plant Dis.* 92(2), 215–24. doi:10.1094/PDIS-92-2-0215.

Francese, J. A., Crook, D. J., Fraser, I., Lance, D. R., Sawyer, A. J. and Mastro, V. C. 2010a. Optimization of trap color for emerald ash borer (Coleoptera: Buprestidae). *J. Econ. Entomol.* 103(4), 1235–41. doi:10.1603/ec10088.

Francese, J. A., Fraser, I., Rietz, M. L., Crook, D. J., Lance, D. R. and Mastro, V. C. 2010b. Relation of color, size, and canopy placement of prism traps in determining capture of emerald ash borer (Coleoptera: Buprestidae). *Can. Entomol.* 142(6), 596–600. doi:10.4039/n10-041.

Francese, J. A., Fraser, I., Lance, D. R. and Mastro, V. C. 2011. Efficacy of multifunnel traps for capturing emerald ash borer (Coleoptera: Buprestidae): effect of color, glue, and other trap coatings. *J. Econ. Entomol.* 104(3), 901–8. doi:10.1603/ec11038.

Freyman, T. 2015. The monitoring of Asian gypsy moth, pink gypsy moth and nun moth at the ports of Vladivostok (Russky Island), Nakhodka, Votochny, Slavyanka, Olga, Vanino, Plastun, Pos'et, Zarubino, Kozmino, Korsakov in 2015. OFO-PR-15-01. Federal Service for Veterinary and Phytosanitary Surveillance, Federal State Budgetary Institution, All-Russia Plant Quarantine Service, Ogden, UT, 51p.

Fuentealba, A., Alfaro, R. and Bauce, É. 2013. Theoretical framework for assessment of risks posed to Canadian forests by invasive insect species. *For. Ecol. Manag.* 302, 97–106. doi:10.1016/j.foreco.2013.03.023.

Fuentealba, A., Dupont, A., Hébert, C., Berthiaume, R., Quezada-Garcia, R. and Bauce, É. 2019. Comparing the efficacy of various aerial spraying scenarios using *Bacillus thuringiensis* to protect trees from spruce budworm defoliation. *For. Ecol. Manag.* 432, 1013–21. doi:10.1016/j.foreco.2018.10.034.

Fuester, R., Hajek, A. E., Schaefer, P. and Elkinton, J. S. 2014. Biological control of *Lymantria dispar*. In: Van Driesche, R. and Reardon, R. (Eds), *The Use of Classical Biological Control to Preserve Forests in North America*. FHTET-2013-02. USDA Forest Service, Washington, DC, pp. 49-82.

Gandhi, K. J. K. and Herms, D. A. 2010. Potential biodiversity loss due to impending devastation of the North American genus *Fraxinus* by the exotic emerald ash borer. *Biol. Invas.* 12(6), 1839–46. doi:10.1007/s10530-009-9594-1.

Garnas, J. R. 2018. Rapid evolution of insects to global environmental change: conceptual issues and empirical gaps. *Curr. Opin. Insect Sci.* 29, 93–101. doi:10.1016/j.cois.2018.07.013.

Gauthier, S., Bernier, P., Kuuluvainen, T., Shvidenko, A. Z. and Schepaschenko, D. G. 2015. Boreal forest health and global change. *Science* 349(6250), 819–22. doi:10.1126/science.aaa9092.

Gent, C. A., Wainhouse, D., Day, K. R., Peace, A. J. and Inward, D. J. G. 2017. Temperature-dependent development of the great European spruce bark beetle *Dendroctonus micans* (Kug.) (Coleoptera: Curculionidae: Scolytinae) and its predator *Rhizophagus grandis* Gyll. (Coleoptera: Monotomidae: Rhizophaginae). *Agric. For. Entomol.* 19(3), 321–31. doi:10.1111/afe.12212.

Gienapp, P., Teplitsky, C., Alho, J. S., Mills, J. A. and Merilä, J. 2008. Climate change and evolution: disentangling environmental and genetic responses. *Mol. Ecol.* 17(1), 167–78. doi:10.1111/j.1365-294X.2007.03413.x.

Gillette, N. E., Erbilgin, N., Webster, J. N., Pederson, L., Mori, S. R., Stein, J. D., Owen, D. R., Bischel, K. M. and Wood, D. L. 2009. Aerially applied verbenone-releasing laminated flakes protect *Pinus contorta* stands from attack by *Dendroctonus ponderosae* in California and Idaho. *For. Ecol. Manag.* 257(5), 1405–12. doi:10.1016/j.foreco.2008.12.017.

Gillette, N. E., Hansen, E. M., Mehmel, C. J., Mori, S. R., Webster, J. N., Erbilgin, N. and Wood, D. L. 2012. Area-wide application of verbenone-releasing flakes reduces mortality of whitebark pine *Pinus albicaulis* caused by the mountain pine beetle *Dendroctonus ponderosae*. *Agric. For. Entomol.* 14(4), 367–75. doi:10.1111/j.1461-9563.2012.00577.x.

Gmelin, J. F. 1787. *Abhandlung über Die Wurmtrocknis*. Cruisiusschen, Leipzig.

Gomez, D. F., Skelton, J., Steininger, M. S., Stouthamer, R., Rugman-Jones, P., Sittichaya, W., Rabaglia, R. J. and Hulcr, J. 2018. Species delineation within the *Euwallacea fornicatus* (Coleoptera: Curculionidae) complex revealed by morphometric and phylogenetic analyses. *Insect Syst. Divers.* 2(6). doi:10.1093/isd/ixy018.

Göthlin, E., Schroeder, L. M. and Lindelöw, A. 2000. Attacks by *Ips typographus* and *Pityogenes chalcographus* on windthrown spruces (*Picea abies*) during the two years following a storm felling. *Scan. J. For. Res.* 15(5), 542–9. doi:10.1080/028275800750173492.

Gray, D. R. 2008. The relationship between climate and outbreak characteristics of the spruce budworm in eastern Canada. *Clim. Change* 87(3–4), 361–83. doi:10.1007/s10584-007-9317-5.

Gray, D. R. 2017. Risk analysis of the invasion pathway of the Asian gypsy moth: a known forest invader. *Biol. Invas.* 19(11), 3259–72. doi:10.1007/s10530-017-1425-1.

Grayson, K. L. and Johnson, D. M. 2018. Novel insights on population and range edge dynamics using an unparalleled spatiotemporal record of species invasion. *J. Anim. Ecol.* 87(3), 581–93. doi:10.1111/1365-2656.12755.

Grosman, D. M., Fettig, C. J., Jorgensen, C. L. and Munson, A. S. 2010. Effectiveness of two systemic insecticides for protecting western conifers from mortality due to bark beetle attack. *West. J. Appl. For.* 25, 181–85.

Haavik, L. J., Dodds, K. J. and Allison, J. D. 2015. Do native insects and associated fungi limit non-native woodwasp, *Sirex noctilio*, survival in a newly invaded environment? *PLoS ONE* 10(10), e0138516. doi:10.1371/journal.pone.0138516.

Hajek, A. E. and van Nouhuys, S. 2016. Fatal diseases and parasitoids: from competition to facilitation in a shared host. *Proc. R. Soc. B: Biol. Sci.* 283(1828). doi:10.1098/rspb.2016.0154.

Hanks, L., Campbell, C., Paine, T. and Millar, J. 1997. Host range expansion of *Helcostizus rufiscutum* Cushman (Hymenoptera: Ichneumonidae) to *Phoracantha semipunctata* Fabr. (Coleoptera: Cerambycidae) in California. *Pan-Pac. Entomol.* 73, 190–1.

Hansen, E. M., Amacher, M. C., Van Miegroet, H., Long, J. N. and Ryan, M. G. 2015. Carbon dynamics in central US Rockies lodgepole pine type after mountain pine beetle outbreaks. *For. Sci.* 61(4), 665–79. doi:10.5849/forsci.14-094.

Harper, A. L., McKinney, L. V., Nielsen, L. R., Havlickova, L., Li, Y., Trick, M., Fraser, F., Wang, L., Fellgett, A., Sollars, E. S., Janacek, S. H., Downie, J. A., Buggs, R. J., Kjær, E. D. and Bancroft, I. 2016. Molecular markers for tolerance of European ash (*Fraxinus excelsior*) to dieback disease identified using associative transcriptomics. *Sci. Rep.* 6, 19335. doi:10.1038/srep19335.

Harrison, R. L., Rowley, D. L. and Keena, M. A. 2016. Geographic isolates of *Lymantria dispar* multiple nucleopolyhedrovirus: genome sequence analysis and pathogenicity against European and Asian gypsy moth strains. *J. Invertebr. Pathol.* 137, 10-22. doi:10.1016/j.jip.2016.03.014.

Hart, S. J., Schoennagel, T., Veblen, T. T. and Chapman, T. B. 2015. Area burned in the western United States is unaffected by recent mountain pine beetle outbreaks. *Proc. Natl. Acad. Sci. U.S.A.* 112(14), 4375-80. doi:10.1073/pnas.1424037112.

Hart, S. J., Veblen, T. T., Schneider, D. and Molotch, N. P. 2017. Summer and winter drought drive the initiation and spread of spruce beetle outbreak. *Ecology* 98(10), 2698-707. doi:10.1002/ecy.1963.

Haynes, K. J., Allstadt, A. J. and Klimetzek, D. 2014. Forest defoliator outbreaks under climate change: effects on the frequency and severity of outbreaks of five pine insect pests. *Glob. Change Biol.* 20(6), 2004-18. doi:10.1111/gcb.12506.

He, S. Y., Ge, X. Z., Wang, T., Wen, J. B. and Zong, S. X. 2015. Areas of potential suitability and survival of *Dendroctonus valens* in China under extreme climate warming scenario. *Bull. Entomol. Res.* 105(4), 477-84. doi:10.1017/S0007485315000309.

Hébert, C., Jobin, L., Berthiaume, R., Coulombe, C. and Dupont, A. 2001. Changes in hemlock looper [Lepidoptera: Geometridae] pupal distribution through a 3-year outbreak cycle. *Phytoprotection* 82(2), 57-63. doi:10.7202/706216ar.

Heidger, C. M. and Lieutier, F. 2002. Possibilities to utilize tree resistance to insects in forest pest management in central and western Europe. In: Wagner, M. R., Clancy, K. M., Lieutier, F. and Paine, T. D. (Eds), *Mechanisms and Deployment of Resistance in Trees to Insects*. Kluwer Academic Publishers, Dordrecht, pp. 239-63. Chapter 11.

Hérard, F. and Maspero, M. 2019. History of discoveries and management of the citrus longhorned beetle, *Anoplophora chinensis*, in Europe. *J. Pest Sci.* 92, 117-30.

Herms, D. A. and McCullough, D. G. 2014. Emerald ash borer invasion of North America: history, biology, ecology, impacts, and management. *Annu. Rev. Entomol.* 59, 13-30. doi:10.1146/annurev-ento-011613-162051.

Hicke, J. A., Johnson, M. C., Hayes, J. L. and Preisler, H. K. 2012. Effects of bark beetle-caused tree mortality on wildfire. *For. Ecol. Manag.* 271, 81-90. doi:10.1016/j.foreco.2012.02.005.

Hofstetter, R. W., Aflitto, N., Bedoya, C. L., Yturralde, K. and Dunn, D. D. 2019. Vibrational behavior in bark beetles: applied aspects. In: Hill, P. S. M., Lakes-Harlan, R., Mazzoni, V., Narins, P., Virant-Doberlet, M. and Wessels, A. (Eds), *Biotremology: Studying Vibrational Behavior, Animal Signals and Communications* (6). Springer, Berlin Heidelberg. Chapter 21.

Houston, D. R. 1994. Major new tree disease epidemics: beech bark disease. *Annu. Rev. Phytopathol.* 32(1), 75-87. doi:10.1146/annurev.py.32.090194.000451.

Hu, J., Angeli, S., Schuetz, S., Luo, Y. and Hajek, A. E. 2009. Ecology and management of exotic and endemic Asian longhorned beetle *Anoplophora glabripennis*. *Agric. Forest Entomol.* 11(4), 359-75. doi:10.1111/j.1461-9563.2009.00443.x.

Hudgins, E. J., Liebhold, A. M. and Leung, B. 2017. Predicting the spread of all invasive forest pests in the United States. *Ecol. Lett.* 20(4), 426-35. doi:10.1111/ele.12741.

Hughes, M. A., Smith, J. A., Ploetz, R. C., Kendra, P. E., Mayfield, A. E., Hanula, J. L., Hulcr, J., Stelinski, L. L., Cameron, S., Riggins, J. J., Carrillo, D., Rabaglia, R., Eickwort, J. and

Pernas, T. 2015. Recovery plan for laurel wilt on redbay and other forest species caused by *Raffaelea lauricola* and disseminated by *Xyleborus glabratus*. *Plant Health Prog.* 16(4), 173–210. doi:10.1094/PHP-RP-15-0017.

Hulcr, J. and Dunn, R. R. 2011. The sudden emergence of pathogenicity in insect-fungus symbioses threatens naive forest ecosystems. *Proc. R. Soc. B: Biol. Sci.* 278(1720), 2866–73. doi:10.1098/rspb.2011.1130.

Hulcr, J., Black, A., Prior, K., Chen, C.-Y. and Li, H.-F. 2017. Studies of ambrosia beetles (Coleoptera: Curculionidae) in their native ranges help predict invasion impact. *Fla. Entomol.* 100(2), 257–61. doi:10.1653/024.100.0219.

Humble, L. M., Mastro, V. and Munson, A. S. 2013. Asian gypsy moth, it's back! In: McManus, K. A. and Gottschalk, K. W. (Eds), *Proceedings of the 24th USDA Interagency Research Forum on Invasive Species*, 8–11 January 2013, Annapolis, MD. Publication FHTET-13-01. US Dept. of Agriculture, Forest Service, Forest Health Technology Enterprise Team, Fort Collins, CO, 124p.

Hurley, B. P., Garnas, J., Wingfield, M. J., Branco, M., Richardson, D. M. and Slippers, B. 2016. Increasing numbers and intercontinental spread of invasive insects on eucalypts. *Biol. Invas.* 18(4), 921–33. doi:10.1007/s10530-016-1081-x.

IPCC. 2014. Climate change 2014: synthesis report. Contribution of working groups I, II and III to the fifth assessment report of the Intergovernmental Panel on Climate Change [Core Writing Team, Pachauri, R. K. and Meyer, L. A. (Eds)]. IPCC, Geneva, Switzerland, 151pp.

Ireland, K. B., Bulman, L., Hoskins, A. J., Pinkard, E. A., Mohammed, C. and Kriticos, D. J. 2018. Estimating the potential geographical range of *Sirex noctilio*: comparison with an existing model and relationship with field severity. *Biol. Invas.* 20(9), 2599–622. doi:10.1007/s10530-018-1721-4.

Ishikawa, T. and Kikuhara, Y. 2009. *Leptoglossus occidentalis* Heidemann (Hemiptera: Coreidae), a presumable recent invader to Japan. *Jpn. J. Entomol.* 12, 115-6.

Islam, M. S., Barr, N. B., Braswell, W. E., Martinez, M., Ledezma, L. A., Molongoski, J., Mastro, V. and Schuenzel, E. L. 2015. A multiplex real-time PCR assay for screening gypsy moths (Lepidoptera: Erebidae) in the United States for evidence of an Asian genotype. *J. Econ. Entomol.* 108(5), 2450–7. doi:10.1093/jee/tov212.

Jacobi, W. R., Koski, R. D. and Negrón, J. F. 2013. Dutch elm disease pathogen transmission by the banded elm bark beetle *Scolytus schevyrewi*. *For. Path.* 43(3), 232–7. doi:10.1111/efp.12023.

Jactel, H., Petit, J., Desprez-Loustau, M. L., Delzon, S., Piou, D., Battisti, A. and Koricheva, J. 2012. Drought effects on damage by forest insects and pathogens: a meta-analysis. *Glob. Change Biol.* 18(1), 267–76. doi:10.1111/j.1365-2486.2011.02512.x.

Jactel, H., Bauhus, J., Boberg, J., Bonal, D., Castagneyrol, B., Gardiner, B., Gonzalez-Olabarria, J. R., Koricheva, J., Meurisse, N. and Brockerhoff, E. G. 2017. Tree diversity drives forest stand resistance to natural disturbances. *Curr. For. Rep.* 3(3), 223–43. doi:10.1007/s40725-017-0064-1.

Janes, J. K., Li, Y., Keeling, C. I., Yuen, M. M., Boone, C. K., Cooke, J. E., Bohlmann, J., Huber, D. P., Murray, B. W., Coltman, D. W. and Sperling, F. A. 2014. How the mountain pine beetle (*Dendroctonus ponderosae*) breached the Canadian Rocky Mountains. *Mol. Biol. Evol.* 31(7), 1803–15. doi:10.1093/molbev/msu135.

Jenkins, M. J., Hebertson, E., Page, W. G. and Jorgensen, C. A. 2008. Bark beetles, fuels, fires and implications for forest management in the Intermountain West. *For. Ecol. Manag.* 254(1), 16–34. doi:10.1016/j.foreco.2007.09.045.

Jenkins, M. J., Page, W. G., Hebertson, E. G. and Alexander, M. E. 2012. Fuels and fire behavior dynamics in bark beetle attacked forests in western North America and implications for fire management. *For. Ecol. Manag.* 275, 23–34. doi:10.1016/j.foreco.2012.02.036.

Jenkins, M. J., Runyon, J. B., Fettig, C. J., Page, W. G. and Bentz, B. J. 2014. Interactions among the mountain pine beetle, fires, and fuels. *For. Sci.* 60(3), 489–501. doi:10.5849/forsci.13-017.

Jennings, D. E., Duan, J. J., Bean, D., Gould, J. R., Rice, K. A. and Shrewsbury, P. M. 2016. Monitoring the establishment and abundance of introduced parasitoids of emerald ash borer larvae in Maryland, USA. *Biol. Control* 101, 138–44. doi:10.1016/j.biocontrol.2016.07.006.

Jepsen, J. U., Hagen, S. B., Ims, R. A. and Yoccoz, N. G. 2008. Climate change and outbreaks of the geometrids *Operophtera brumata* and *Epirrita autumnata* in subarctic birch forest: evidence of a recent outbreak range expansion. *J. Anim. Ecol.* 77(2), 257–64. doi:10.1111/j.1365-2656.2007.01339.x.

Jepsen, J. U., Kapari, L., Hagen, S. B., Schott, T., Vindstad, O. P. L., Nilssen, A. C. and Ims, R. A. 2011. Rapid northwards expansion of a forest insect pest attributed to spring phenology matching with sub-Arctic birch. *Glob. Change Biol.* 17(6), 2071–83. doi:10.1111/j.1365-2486.2010.02370.x.

Johansson, A., Birgersson, G. and Schlyter, F. 2019. Using synthetic semiochemicals to train canines to detect bark beetle-infested trees. *Ann. For. Sci.* 76(2). doi:10.1007/s13595-019-0841-z.

Johnson, D. M., Liebhold, A. M., Tobin, P. C. and Bjørnstad, O. N. 2006. Allee effects and pulsed invasion by the gypsy moth. *Nature* 444(7117), 361–3. doi:10.1038/nature05242.

Johnson, D. M., Büntgen, U., Frank, D. C., Kausrud, K., Haynes, K. J., Liebhold, A. M., Esper, J. and Stenseth, N. C. 2010. Climatic warming disrupts recurrent Alpine insect outbreaks. *Proc. Natl. Acad. Sci. U.S.A.* 107(47), 20576–81. doi:10.1073/pnas.1010270107.

Jönsson, A. M., Appelberg, G., Harding, S. and Bärring, L. 2009. Spatio-temporal impact of climate change on the activity and voltinism of the spruce bark beetle, *Ips typographus*. *Glob. Change Biol.* 15(2), 486–99. doi:10.1111/j.1365-2486.2008.01742.x.

Jönsson, A. M., Harding, S., Krokene, P., Lange, H., Lindelöw, Å., Økland, B., Ravn, H. P. and Schroeder, L. M. 2011. Modelling the potential impact of global warming on *Ips typographus* voltinism and reproductive diapause. *Clim. Change* 109(3–4), 695–718. doi:10.1007/s10584-011-0038-4.

Kane, J. M., Varner, J. M., Metz, M. R. and van Mantgem, P. J. 2017. Characterizing interactions between fire and other disturbances and their impacts on tree mortality in western US Forests. *For. Ecol. Manag.* 405, 188–99. doi:10.1016/j.foreco.2017.09.037.

Kautz, M., John, R., Delb, H. and Puhlmann, H. 2018. Borkenkäfer-Monitoring 2.0: Online-Tool zur flächendeckenden Abschätzung des aktuellen Befallsrisikos durch den Buchdrucker. Berichte des Forschungszentrums Waldökosysteme der Universität Göttingen, Band 83, 351p.

Keane, R. M. and Crawley, M. J. 2002. Exotic plant invasions and the enemy release hypothesis. *Trends Ecol. Evol.* 17(4), 164–70. doi:10.1016/S0169-5347(02)02499-0.

Keena, M. A. 2016. Inheritance and world variation in thermal requirements for egg hatch in *Lymantria dispar* (Lepidoptera: Erebidae). *Environ. Entomol.* 45(1), 1–10. doi:10.1093/ee/nvv163.

Kenis, M., Hurley, B. P., Hajek, A. E. and Cock, M. J. W. 2017. Classical biological control of insect pests of trees: facts and figures. *Biol. Invas.* 19(11), 3401–17. doi:10.1007/s10530-017-1414-4.

Kharuk, V. I., Im, S. T. and Yagunov, M. N. 2018. Migration of the northern boundary of the Siberian silk moth. *Contemp. Probl. Ecol.* 11(1), 26–34. doi:10.1134/S1995425518010055.

King, J. N., Alfaro, R. I., Lopez, M. G. and Van Akker, L. 2011. Resistance of Sitka spruce (*Picea sitchensis* (Bong.) Carr.) to white pine weevil (*Pissodes strobi* Peck): characterizing the bark defence mechanisms of resistant populations. *Forestry* 84(1), 83–91. doi:10.1093/forestry/cpq047.

Kirichenko, N. I., Baranchikov, Y. N. and Vidal, S. 2009. Performance of the potentially invasive Siberian moth *Dendrolimus superans sibiricus* on coniferous species in Europe. *Ag. For. Entomol.* 11(3), 247–54. doi:10.1111/j.1461-9563.2009.00437.x.

Kiss, G. K. and Yanchuk, A. D. 1991. Preliminary evaluation of genetic variation of weevil resistance in interior spruce in British Columbia. *Can. J. For. Res.* 21(2), 230–4. doi:10.1139/x91-028.

Koch, J. L., Carey, D. W., Mason, M. E., Poland, T. M. and Knight, K. S. 2015. Intraspecific variation in *Fraxinus pennsylvanica* responses to emerald ash borer (*Agrilus planipennis*). *New Forests* 46(5–6), 995–1011. doi:10.1007/s11056-015-9494-4.

Köhl, M., Lasco, R., Cifuentes, M., Jonsson, Ö., Korhonen, K. T., Mundhenk, P., de Jesus Navar, J. and Stinson, G. 2015. Changes in forest production, biomass and carbon: results from the 2015 UN FAO Global Forest Resource Assessment. *For. Ecol. Manag.* 352, 21–34. doi:10.1016/j.foreco.2015.05.036.

Kolb, T. E., Fettig, C. J., Ayres, M. P., Bentz, B. J., Hicke, J. A., Mathiasen, R., Stewart, J. E. and Weed, A. S. 2016. Observed and anticipated impacts of drought on forest insects and diseases in the United States. *For. Ecol. Manag.* 380, 321–34. doi:10.1016/j.foreco.2016.04.051.

Kölling, C., Knoke, T., Schall, P. and Ammer, C. 2009. Überlegungen zum Risiko des Fichtenanbaus in Deutschland vor dem Hintergrund des Klimawandels [Cultivation of Norway spruce (*Picea abies* (L.) Karst.) in Germany: considerations on risk against the background of climate change]. *Forstarchiv* 80, 42–54.

Kononov, A., Ustyantsev, K., Wang, B., Mastro, V. C., Fet, V., Blinov, A. and Baranchikov, Y. 2016. Genetic diversity among eight *Dendrolimus* species in Eurasia (Lepidoptera: Lasiocampidae) inferred from mitochondrial COI and COII, and nuclear ITS2 markers. *BMC Genet.* 17(Suppl. 3), 157. doi:10.1186/s12863-016-0463-5.

Koontz, M. J., Latimer, A. M., Mortenson, L. A., Fettig, C. J., Millar, C. I. and North, M. P. 2018. The effect of spatial variability of forest structure on severity of a tree-killing insect. In: *MtnClim 2018: 8th Mountain Climate Conference*, 17–21 September 2018, Gothic, CO.

Kovacs, K. F., Haight, R. G., McCullough, D. G., Mercader, R. J., Siegert, N. W. and Liebhold, A. M. 2010. Cost of potential emerald ash borer damage in U.S. communities, 2009-2019. *Ecol. Econ.* 69(3), 569–78. doi:10.1016/j.ecolecon.2009.09.004.

Kreutz, J., Zimmermann, G. and Vaupel, O. 2010. Horizontal transmission of the entomopathogenic fungus *Beauveria bassiana* among the spruce bark beetle, *Ips typographus* (Col., Scolytidae) in the laboratory and under field conditions. *Biocontrol Sci. Tech.* 14, 837–48.

Langor, D. W., DeHass, L. J. and Footit, R. G. 2009. Diversity of non-native terrestrial arthropods on woody plants in Canada. *Biol. Invas.* 11(1), 5–19. doi:10.1007/s10530-008-9327-x.

Lantschner, M. V., Villacide, J. M., Garnas, J. R., Croft, P., Carnegie, A. J., Liebhold, A. M. and Corley, J. C. 2014. Temperature explains variable spread rates of the invasive woodwasp *Sirex noctilio* in the Southern Hemisphere. *Biol. Invas.* 16(2), 329–39. doi:10.1007/s10530-013-0521-0.

Lantschner, M. V., Atkinson, T. H., Corley, J. C. and Liebhold, A. M. 2017. Predicting North American Scolytinae invasions in the Southern Hemisphere. *Ecol. App.* 27(1), 66–77. doi:10.1002/eap.1451.

Lantschner, M. V., Aukema, B. H. and Corley, J. C. 2019. Droughts drive outbreak dynamics of an invasive forest insect on an exotic host. *For. Ecol. Manag.* 433, 762–70. doi:10.1016/j.foreco.2018.11.044.

Lapointe, R., Thumbi, D. and Lucarotti, C. J. 2012. Recent advances in our knowledge of baculovirus molecular biology and its relevance for the registration of baculovirus-based products for insect pest population control. In: Soloneski, S. (Ed.), *Integrated Pest Management and Pest Control-Current and Future Tactics*. IntechOpen.com, Rijeka, pp. 481–552. Chapter 21.

Lee, H. R., Lee, S. C., Lee, D. H., Choi, W. S., Jung, C. S., Jeon, J. H., Kim, J. E. and Park, I. K. 2017. Identification of the aggregation-sex pheromone produced by male *Monochamus saltuarius*, a major insect vector of the pine wood nematode. *J. Chem. Ecol.* 43(7), 670–8. doi:10.1007/s10886-017-0864-6.

Leonard, D. S. 2007. Project organization, planning and operations. In: Tobin, P. C. and Blackburn, L. M. (Eds), *Slow the Spread: A National Program to Manage the Gypsy Moth*. USDA Forest Service Northern Research Station General Technical Report NRS-6, pp. 91–9.

Leonard, D. S. and Sharov, A. A. 1995. Slow the Spread project update: developing a process for evaluation. In: *Proceedings. U.S. Dep. Agric. Interagency Gypsy Moth Research Forum 1995*. USDA For. Serv. Gen. Tech. Rep. NE-213, pp. 82–5.

Leonhardt, B. A., Mastro, V. C., Leonard, D. S., McLane, W., Reardon, R. C. and Thorpe, K. W. 1996. Control of low-density gypsy moth (Lepidoptera: Lymantriidae) populations by mating disruption with pheromone. *J. Chem. Ecol.* 22(7), 1255–72. doi:10.1007/BF02266964.

Lesieur, V., Lombaert, E., Guillemaud, T., Courtial, B., Strong, W., Roques, A. and Auger-Rozenberg, M.-A. 2019. The rapid spread of *Leptoglossus occidentalis* in Europe: a bridgehead invasion. *J. Pest. Sci.* 92(1), 189–200. doi:10.1007/s10340-018-0993-x.

Lesk, C., Coffel, E., D'Amato, A. W., Dodds, K. and Horton, R. 2017. Threats to North American forests from southern pine beetle with warming winters. *Nat. Clim. Change* 7(10), 713–7. doi:10.1038/nclimate3375.

Leuschner, W. A., Young, J. A., Walden, S. A. and Ravlin, F. W. 1996. Potential benefits of slowing the gypsy moth's spread. *South. J. App. For.* 20, 65–73.

Lewis, E. E. and Cane, J. H. 1990. Stridulation as a primary anti-predator defence of a beetle. *Anim. Behav.* 40(5), 1003–4. doi:10.1016/S0003-3472(05)81011-5.

Liebhold, A. M. 1992. Are North American populations of gypsy moth (Lepidoptera: Lymantriidae) bimodal? *Environ. Entomol.* 21(2), 221-9. doi:10.1093/ee/21.2.221.

Liebhold, A. M. and McManus, M. 1999. The evolving use of insecticides in gypsy moth management. *J. For.* 97, 20-3.

Liebhold, A. M. and Tobin, P. C. 2006. Growth of newly established alien populations: comparison of North American gypsy moth colonies with invasion theory. *Pop. Ecol.* 48(4), 253-62. doi:10.1007/s10144-006-0014-4.

Liebhold, A. M. and Tobin, P. C. 2008. Population ecology of insect invasions and their management. *Annu. Rev. Entomol.* 53, 387-408. doi:10.1146/annurev.ento.52.110405.091401.

Liebhold, A. M., Mastro, V. and Schaefer, P. W. 1989. Learning from the legacy of Leopold Trouvelot. *Bull. Entomol. Soc. Am.* 35(2), 20-2. doi:10.1093/besa/35.2.20.

Liebhold, A. M., Twardus, D. and Buonaccorsi, J. 1991. Evaluation of the timed-walk method of estimating gypsy moth, *Lymantria dispar* (Lepidoptera: Lymantriidae) egg mass densities. *J. Econ. Entomol.* 84(6), 1774-81. doi:10.1093/jee/84.6.1774.

Liebhold, A. M., MacDonald, W. L., Bergdahl, D. and Mastro, V. C. 1995. Invasion by exotic forest pests: a threat to forest ecosystems. *For. Sci.* 41(suppl_1), 1-49. doi:10.1093/forestscience/41.s1.a0001.

Liebhold, A. M., Luzader, E., Reardon, R., Bullard, A., Roberts, A., Ravlin, W., Delost, S. and Spears, B. 1996. Use of a geographic information system to evaluate regional treatment effects in a gypsy moth (Lepidoptera: Lymantriidae) management program. *J. Econ. Entomol.* 89(5), 1192-203. doi:10.1093/jee/89.5.1192.

Liebhold, A. M., Luzader, E., Reardon, R., Roberts, A., Ravlin, W. F., Sharov, A. and Zhou, G. 1998. Forecasting gypsy moth (Lepidoptera: Lymantriidae) defoliation with a geographical information system. *J. Econ. Entomol.* 91(2), 464-72. doi:10.1093/jee/91.2.464.

Liebhold, A. M., Elkinton, J., Williams, D. and Muzika, R. M. 2000. What causes outbreaks of the gypsy moth in North America? *Pop. Ecol.* 42(3), 257-66. doi:10.1007/PL00012004.

Liebhold, A. M., Sharov, A. A. and Tobin, P. C. 2007. Population biology of gypsy moth spread. In: Tobin, P. C. and Blackburn, L. M. (Eds), *Slow the Spread: A National Program to Manage the Gypsy Moth*. USDA Forest Service Northern Research Station General Technical Report NRS-6, pp. 15-32.

Liebhold, A. M., Brockerhoff, E. G., Kalisz, S., Nuñez, M. A., Wardle, D. A. and Wingfield, M. J. 2017. Biological invasions in forest ecosystems. *Biol. Invas.* 19(11), 3437-58. doi:10.1007/s10530-017-1458-5.

Logan, J. A., Régnière, J. and Powell, J. A. 2003. Assessing the impacts of global warming on forest pest dynamics. *Front. Ecol. Environ.* 1(3), 130-7. doi:10.1890/1540-9295(2003)001[0130:ATIOGW]2.0.CO;2.

Lombardo, J. A. and Elkinton, J. S. 2017. Environmental adaptation in an asexual invasive insect. *Ecol. Evol.* 7(14), 5123-30. doi:10.1002/ece3.2894.

López-Goldar, X., Villari, C., Bonello, P., Borg-Karlson, A. K., Grivet, D., Zas, R. and Sampedro, L. 2018. Inducibility of plant secondary metabolites in the stem predicts genetic variation in resistance against a key insect herbivore in maritime pine. *Front. Plant Sci.* 9, 1651. doi:10.3389/fpls.2018.01651.

Lu, M. and Sun, J. 2017. Biological invasions in forest ecosystem in China. In: Wan, F., Jiang, M. and Zhan, A. (Eds), *Biological Invasions and Its Management in China*. Springer, Dordecht, pp. 53-66.

Lynch, S. C., Twizeyimana, M., Mayorquin, J. S., Wang, D. H., Na, F., Kayim, M., Kasson, M. T., Thu, P. Q., Bateman, C., Rugman-Jones, P., Hulcr, J., Stouthamer, R. and Eskalen, A. 2016. Identification, pathogenicity and abundance of *Paracremonium pembeum* sp. nov. and *Graphium euwallaceae* sp. nov.–two newly discovered mycangial associates of the polyphagous shot hole borer (*Euwallacea* sp.) in California. *Mycologia* 108(2), 313–29. doi:10.3852/15-063.

Maclauchlan, L. E., Daniels, L. D., Hodge, J. C. and Brooks, J. E. 2018. Characterization of western spruce budworm outbreak regions in the British Columbia Interior. *Can. J. For. Res.* 48, 783–802.

MacLean, D. A., Beaton, K. P., Porter, K. B., MacKinnon, W. E. and Budd, M. G. 2002. Potential wood supply losses to spruce budworm in New Brunswick estimated using the Spruce Budworm Decision Support System. *For. Chron.* 78(5), 739–50. doi:10.5558/tfc78739-5.

MacQuarrie, C. J. K., Lyons, D. B., Seehausen, M. L. and Smith, S. M. 2016. A history of biological control in Canadian forests, 1882–2014. *Can. Entomol.* 148(S1), S239–69. doi:10.4039/tce.2015.66.

Marini, L., Økland, B., Jönsson, A. M., Bentz, B., Carroll, A., Forster, B., Grégoire, J., Hurling, R., Nageleisen, L. M., Netherer, S., Ravn, H. P., Weed, A. and Schroeder, M. 2017. Climate drivers of bark beetle outbreak dynamics in Norway spruce forests. *Ecography* 40(12), 1426–35. doi:10.1111/ecog.02769.

Martel, V., Johns, R., Eveleigh, E., McCann, K., Pureswaran, D., Sylvain, Z., Morrison, A., Morin, B., Owens, E. and Hébert, C. 2018. Landscape level impacts of early intervention strategy on spruce budworm, other herbivores and associated natural enemies. In: *SERG-International Workshop Proceedings*, Edmonton, AB, pp. 138–46.

Mattor, K. M., Morris, J. L., Fettig, C. J., Cottrell, S., McGrady, P., Maguire, D., James, P. M. A., Clear, J., Wurtzebach, Z., Wei, Y., Brunelle, A., Western, J., Maxwell, R., Rotar, M., Gallagher, L. and Roberts, R. 2019. Adaptive capacity in social-ecological systems: a framework for bark beetle disturbances in natural resource management. *Sustain. Sci.* (in review).

Mattson, W. J. and Addy, N. D. 1975. Phytophagous insects as regulators of forest primary production. *Science* 190, 515–22.

McAvoy, T., Régnière, J., St-Amant, R., Schneeberger, N. and Salom, S. 2017. Mortality and recovery of hemlock woolly adelgid (*Adelges tsugae*) in response to winter temperatures and predictions for the future. *Forests* 8(12), 497. doi:10.3390/f8120497.

McCullough, D. G. and Mercader, R. J. 2012. SLAM in an urban forest: evaluation of potential strategies to SLow Ash Mortality caused by emerald ash borer (*Agrilus planipennis*). *Int. J. Pest Manag.* 58(1), 9–23. doi:10.1080/09670874.2011.637138.

McCullough, D. G., Work, T. T., Cavey, J. F., Liebhold, A. M. and Marshall, D. 2006. Interceptions of nonindigenous plant pests at US ports of entry and border crossings over a 17-year period. *Biol. Invas.* 8(4), 611–30. doi:10.1007/s10530-005-1798-4.

McCullough, D. G., Poland, T. M., Anulewicz, A. C. and Cappaert, D. 2009a. Emerald ash borer (Coleoptera: Buprestidae) attraction to stressed or baited ash trees. *Environ. Entomol.* 38(6), 1668–79. doi:10.1603/022.038.0620.

McCullough, D. G., Poland, T. M., Cappaert, D. and Anulewicz, A. C. 2009b. Emerald ash borer (*Agrilus planipennis*) attraction to ash trees stressed by girdling, herbicide and wounding. *Can. J. For. Res.* 39(7), 1331–45. doi:10.1139/X09-057.

McCullough, D. G., Poland, T. M., Anulewicz, A. C., Lewis, P. and Cappaert, D. 2011. Evaluation of *Agrilus planipennis* control provided by emamectin benzoate and two neonicotinoid insecticides, one & two seasons after treatment. *J. Econ. Entom.* 104, 1599-612.

McCullough, D. G., Mercader, R. J. and Siegert, N. W. 2015. Developing and integrating tactics to slow ash mortality caused by emerald ash borer. *Can. Entomol.* 147(3), 349-58. doi:10.4039/tce.2015.3.

McKenzie, N., Helson, B., Thompson, D., Otis, G., Mcfarlane, J., Buscarini, T. and Meating, J. 2010. Azadirachtin: an effective systemic insecticide for control of *Agrilus planipennis* (Coleoptera: Buprestidae). *J. Econ. Entomol.* 103(3), 708-17. doi:10.1603/ec09305.

McManis, A. E., Powell, J. A. and Bentz, B. J. 2018. Developmental parameters of a southern mountain pine beetle (Coleoptera: Curculionidae) population reveal potential source of latitudinal differences in generation time. *Can. Entomol.* 151(1), 1-15. doi:10.4039/tce.2018.51.

McManus, M. L. 2007. In the beginning: gypsy moth in the United States. In: Tobin, P. C. and Blackburn, L. M. (Eds), *Slow the Spread: A National Program to Manage the Gypsy Moth*. USDA Forest Service Northern Research Station General Technical Report NRS-6, pp. 3-13.

McManus, M. and Csóka, G. 2007. History and impact of gypsy moth in North America and comparison to the recent outbreaks in Europe. *Acta Silvatica Lignaria Hung.* 3, 47-64.

McPherson, J., Packauskas, R., Taylor, S. and O'Brien, M. 1990. Eastern range extension of *Leptoglossus occidentalis* with a key to *Leptoglossus* species of America north of Mexico (Heteroptera: Coreidae). *Great Lakes Entomol.* 23, 99-104.

Mech, A. M., Tobin, P. C., Teskey, R. O., Rhea, J. R. and Gandhi, K. J. K. 2018. Increases in summer temperatures decrease the survival of an invasive forest insect. *Biol. Invas.* 20(2), 365-74. doi:10.1007/s10530-017-1537-7.

Meigs, G. W., Campbell, J. L., Zald, H. S., Bailey, J. D., Shaw, D. C. and Kennedy, R. E. 2015. Does wildfire likelihood increase following insect outbreaks in conifer forests? *Ecosphere* 6, 1-24.

Mendel, Z., Protasov, A., Fisher, N. and La Salle, J. 2004. Taxonomy and biology of *Leptocybe invasa* gen. & sp. n. (Hymenoptera: Eulophidae), an invasive gall inducer on Eucalyptus. *Aust. J. Entomol.* 43(2), 101-13. doi:10.1111/j.1440-6055.2003.00393.x.

Mendel, Z., Branco, M. and Battisti, A. 2016. Invasive sap-sucker insects in the Mediterranean basin. In: Paine, T. and Lieutier, F. (Eds), *Insects and Diseases of Mediterranean Forest Systems*. Springer, Cham, pp. 261-91.

Mercader, R. J., McCullough, D. G. and Bedford, J. M. 2013. A comparison of girdled ash detection trees and baited artificial traps for emerald ash borer (*Agrilus planipennis* Fairmaire) detection. *Environ. Entomol.* 42(5), 1027-39. doi:10.1603/EN12334.

Mercader, R. J., McCullough, D. G., Storer, A. J., Bedford, J. M., Heyd, R., Poland, T. M. and Katovich, S. 2015. Evaluation of the potential use of a systemic insecticide and girdled trees in area wide management of the emerald ash borer. *For. Ecol. Manag.* 350, 70-80. doi:10.1016/j.foreco.2015.04.020.

Mezei, P., Jakuš, R., Pennerstorfer, J., Havašová, M., Škvarenina, J., Ferenčík, J., Slivinský, J., Bičárová, S., Bilčík, D., Blaženec, M. and Netherer, S. 2017. Storms, temperature maxima and the Eurasian spruce bark beetle *Ips typographus*–an infernal trio in Norway spruce forests of the Central European High Tatra Mountains. *Ag. For. Meteorol.* 242, 85-95. doi:10.1016/j.agrformet.2017.04.004.

Moreira, X., Abdala-Roberts, L., Rasmann, S., Castagneyrol, B. and Mooney, K. A. 2016. Plant diversity effects on insect herbivores and their natural enemies: current thinking, recent findings and future directions. *Curr. Opin. Insect Sci.* 14, 1–7. doi:10.1016/j.cois.2015.10.003.

Morin, R. S. and Liebhold, A. M. 2016. Invasive forest defoliator contributes to the impending downward trend of oak dominance in eastern North America. *Forestry* 89(3), 284–9. doi:10.1093/forestry/cpv053.

Morin, R. S., Liebhold, A. M. and Gottschalk, K. W. 2004. Area-wide analysis of hardwood defoliator effects on tree conditions in the Allegheny Plateau. *North. J. Appl. For.* 21, 31–9.

Morris, J. L., Cottrell, S., Fettig, C. J., Hansen, W. D., Sherriff, R. L., Carter, V. A., Clear, J. L., Clement, J., DeRose, R. J., Hicke, J. A., Higuera, P. E., Mattor, K. M., Seddon, A. W. R., Seppä, H. T., Stednick, J. D. and Seybold, S. J. 2017. Managing bark beetle impacts on ecosystems and society: priority questions to motivate future research. *J. Appl. Ecol.* 54(3), 750–60. doi:10.1111/1365-2664.12782.

Muzika, R. M. 2017. Opportunities for silviculture in management and restoration of forests affected by invasive species. *Biol. Invas.* 19(11), 3419–35. doi:10.1007/s10530-017-1549-3.

Muzika, R. M. and Liebhold, A. M. 1999. Effects of gypsy moth defoliation on radial growth of host and non-host tree species. *Can. J. For. Res.* 29(9), 1365–73. doi:10.1139/x99-098.

Muzika, R. M. and Liebhold, A. M. 2000. A critique of silvicultural approaches to managing defoliating insects in North America. *Ag. For. Entomol.* 2(2), 97–105. doi:10.1046/j.1461-9563.2000.00063.x.

Myers, J. H., Simberloff, D., Kuris, A. M. and Carey, J. R. 2000. Eradication revisited: dealing with exotic species. *Trends Ecol. Evol. (Amst.)* 15(8), 316–20. doi:10.1016/S0169-5347(00)01914-5.

Nakamura, S., Konishi, K. and Takasu, K. 2006. Invasion of the coconut hispine beetle, *Brontispa longissima*: current situation and control measures in Southeast Asia. In: *Proceedings of International Workshop on Development of Database (APASD) for Biological Invasion*. Taiwan Agricultural Chemicals and Toxic Substance Research Institute, pp. 1–9.

Nealis, V. G. 1991. Natural enemies and forest pest management. *For. Chron.* 67(5), 500–5. doi:10.5558/tfc67500-5.

Nealis, V. 2009. Still invasive after all these years: keeping gypsy moth out of British Columbia. *For. Chron.* 85(4), 593–603. doi:10.5558/tfc85593-4.

Negrón, J. F. and Fettig, C. J. 2014. Mountain pine beetle, a major disturbance agent in US western coniferous forests: a synthesis of the state of knowledge. *For. Sci.* 60(3), 409–13. doi:10.5849/forsci.13-169.

Negrón, J. F., Witcosky, J. J., Cain, R. J., LaBonte, J. R., Duerr, D. A., McElwey, S. J., Lee, J. C. and Seybold, S. J. 2005. The banded elm bark beetle: A new threat to the elms in North America. *Am. Entomol.* 51(2), 84–94. doi:10.1093/ae/51.2.84.

Negrón, J. F., McMillin, J. D., Anhold, J. A. and Coulson, D. 2009. Bark beetle-caused mortality in a drought-affected ponderosa pine landscape in Arizona, USA. *For. Ecol. Manag.* 257(4), 1353–62. doi:10.1016/j.foreco.2008.12.002.

Negrón, J. F., Allen, K. K., Ambourn, A., Cook, B. and Marchand, K. 2018. Large-scale thinnings, ponderosa pine, and mountain pine beetle in the Black Hills, USA. *For. Sci.* 63(5), 529–36. doi:10.5849/FS-2016-061.

Netherer, S. and Nopp-Mayr, U. 2005. Predisposition assessment systems (PAS) as supportive tools in forest management-rating of site and stand-related hazards of bark beetle infestation in the High Tatra Mountains as an example for system application and verification. *For. Ecol. Manag.* 207(1-2), 99-107. doi:10.1016/j.foreco.2004.10.020.

Netherer, S. and Schopf, A. 2010. Potential effects of climate change on insect herbivores in European forests–general aspects and the pine processionary moth as specific example. *For. Ecol. Manag.* 259(4), 831-8. doi:10.1016/j.foreco.2009.07.034.

Netherer, S., Matthews, B., Katzensteiner, K., Blackwell, E., Henschke, P., Hietz, P., Pennerstorfer, J., Rosner, S., Kikuta, S., Schume, H. and Schopf, A. 2015. Do water-limiting conditions predispose Norway spruce to bark beetle attack? *New Phytol.* 205(3), 1128-41. doi:10.1111/nph.13166.

Niemelä, P. and Mattson, W. J. 1996. Invasion of North American forests by European phytophagous insects. *BioScience* (46), 741-53.

Økland, B., Netherer, S. and Marini, L. 2015. The Eurasian spruce bark beetle: the role of climate. In: Björkman, C. and Niemelä, P. (Eds), *Climate Change and Insect Pests. CABI Climate Change Series 7*. CABI Publishing, Wallingford, UK, pp. 202-19.

Onufrieva, K. S., Hickman, A. D., Leonard, D. S. and Tobin, P. C. 2019. Relationship between efficacy of mating disruption and gypsy moth density. *Int. J. Pest Manag.* 65(1), 44-52. doi:10.1080/09670874.2018.1455116.

Page, W. G., Jenkins, M. J. and Runyon, J. B. 2012. Mountain pine beetle attack alters the chemistry and flammability of lodgepole pine foliage. *Can. J. For. Res.* 42(8), 1631-47. doi:10.1139/x2012-094.

Paine, T. D. 2016. Insects colonizing eucalypts in California. In: Paine, T. D. and Lieutier, F. (Eds), *Insects and Diseases of Mediterranean Forest Systems*. Springer International Publishing AG, Cham, Switzerland, pp. 711-30.

Pan, Y., Birdsey, R. A., Fang, J., Houghton, R., Kauppi, P. E., Kurz, W. A., Phillips, O. L., Shvidenko, A., Lewis, S. L., Canadell, J. G., Ciais, P., Jackson, R. B., Pacala, S. W., McGuire, A. D., Piao, S., Rautiainen, A., Sitch, S. and Hayes, D. 2011. A large and persistent carbon sink in the world's forests. *Science* 333(6045), 988-93. doi:10.1126/science.1201609.

Parent, G. J., Giguère, I., Germanos, G., Lamara, M., Bauce, É. and MacKay, J. J. 2017. Insect herbivory (*Choristoneura fumiferana*, Tortricidea) underlies tree population structure (*Picea glauca*, Pinaceae). *Sci. Rep.* 7, 42273. doi:10.1038/srep42273.

Peterson, A. T., Williams, R. and Chen, G. 2007. Modeled global invasive potential of Asian gypsy moths, *Lymantria dispar*. *Entomol. Exper. Appl.* 125(1), 39-44. doi:10.1111/j.1570-7458.2007.00603.x.

Pogue, M. and Schaefer, P. 2007. A review of selected species of *Lymantria* Hübner [1819] including three new species (Lepidoptera: Noctuidae: Lymantriinae) from subtropical and temperate regions of Asia, some potentially invasive to North America. Publication FHTET-2006-07. USDA Forest Service Forest Health Technology Enterprise Team, Morgantown, WV, 223pp.

Poland, T. M. and McCullough, D. G. 2006. Emerald ash borer: invasion of the urban forest and the threat to North America's ash resource. *J. For.* 104, 118-24.

Poland, T. M. and McCullough, D. G. 2010. SLAM: a multi-agency pilot project to SLow Ash Mortality caused by emerald ash borer in outlier sites. *Newsl. Mich. Entomol. Soc.* 55(1 and 2), 4-8.

Poland, T. M. and McCullough, D. G. 2014. Comparison of trap types and colors for capturing emerald ash borer adults at different population densities. *Environ. Entomol.* 43(1), 157–70. doi:10.1603/EN13137.

Poland, T. M., McCullough, D. G. and Anulewicz, A. C. 2011. Evaluation of an artificial trap for *Agrilus planipennis* (Coleoptera: Buprestidae) incorporating olfactory and visual cues. *J. Econ. Entomol.* 104(2), 517–31. doi:10.1603/ec10254.

Poland, T. M., Patel-Weynand, T., Finch, D., Ford Miniat, C. and Lopez, V. (Eds). 2019. *National Assessment of Invasive Species*. Springer Verlag, Cham, Switzerland.

Pothier, D., Elie, J.-G., Auger, I., Mailly, D. and Gaudreault, M. 2012. Spruce budworm-caused mortality to balsam fir and black spruce in pure and mixed conifer stands. *For. Sci.* 58(1), 24–33. doi:10.5849/forsci.10-110.

Progar, R. A., Gillette, N., Fettig, C. J. and Hrinkevich, K. 2014. Applied chemical ecology of the mountain pine beetle. *For. Sci.* 60(3), 414–33. doi:10.5849/forsci.13-010.

Pureswaran, D. S., De Grandpré, L., Paré, D., Taylor, A., Barrette, M., Morin, H., Régnière, J. and Kneeshaw, D. D. 2015. Climate-induced changes in host tree-insect phenology may drive ecological state-shift in boreal forests. *Ecology* 96(6), 1480–91. doi:10.1890/13-2366.1.

Pureswaran, D. S., Roques, A. and Battisti, A. 2018. Forest insects and climate change. *Curr. Rep.* 4(2), 35–50. doi:10.1007/s40725-018-0075-6.

Rabaglia, R. J., Cognato, A. I., Hoebeke, E. R., Johnson, C. W., LaBonte, J. R., Carter, M. E. and Vlach, J. J. 2019. Early detection and rapid response: a 10-year summary of the USDA Forest Service program of surveillance for non-native bark and ambrosia beetles. *Am. Entomol.* 65(1), 29–42. doi:10.1093/ae/tmz015.

Raffa, K. F., Aukema, B., Bentz, B. J., Carroll, A., Erbilgin, N., Herms, D., Hicke, J., Hofstetter, R., Katovich, S., Lindgren, B. S., Logan, J., Mattson, W., Munson, A. S., Robison, D. J., Six, D. L., Tobin, P., Townsend, P. and Wallin, K. 2009. A literal use of "forest health" safeguards against misuse and misapplication. *J. For.* 107, 276–7.

Raffa, K. F., Gregoire, J.-C. and Lindgren, B. S. 2015. Natural history and ecology of bark beetles. In: Vega, F. and Hofstetter, R. (Eds), *Bark Beetles, Biology and Ecology of Native and Invasive Species*. Elsevier, London and Oxford, UK, San Diego, CA, Waltham, MA, pp. 1–40. Chapter 1.

Raty, L., Drumont, A., De Windt, N. and Grégoire, J.-C. 1995. Mass trapping of the spruce bark beetle *Ips typographus* L.: traps or trap trees? *For. Ecol. Manag.* 78(1–3), 191–205. doi:10.1016/0378-1127(95)03582-1.

Ratzeburg, J. T. C. 1839. *Die Forst-Insecten. Erster Theil. Die Käfer*. Nicolai'ische Buchhandlung, Berlin, 247p.

Rebek, E. J., Herms, D. A. and Smitley, D. R. 2008. Interspecific variation in resistance to emerald ash borer (Coleoptera: Buprestidae) among North American and Asian ash (*Fraxinus* spp.). *Environ. Entomol.* 37(1), 242–6. doi:10.1603/0046-225X(2008)37[242:IVIRTE]2.0.CO;2.

Régnière, J. and Nealis, V. G. 2007. Ecological mechanisms of population change during outbreaks of the spruce budworm. *Ecol. Entomol.* 32(5), 461–77. doi:10.1111/j.1365-2311.2007.00888.x.

Régnière, J. and Nealis, V. G. 2019. Influence of temperature on historic and future population fitness of the western spruce budworm, *Choristoneura occidentalis*. *Int. J. Pest Manag.*, 1–16. doi:10.1080/09670874.2018.1541113.

Régnière, J., Nealis, V. and Porter, K. 2009. Climate suitability and management of the gypsy moth invasion into Canada. *Biol. Invasions* 11(1), 135–48. doi:10.1007/s10530-008-9325-z.

Régnière, J., St-Amant, R. and Duval, P. 2012. Predicting insect distributions under climate change from physiological responses: spruce budworm as an example. *Biol. Invas.* 14(8), 1571–86. doi:10.1007/s10530-010-9918-1.

Régnière, J., Delisle, J., Pureswaran, D. S. and Trudel, R. 2013. Mate-finding allee effect in spruce budworm population dynamics. *Entomol. Exp. Appl.* 146(1), 112-22. doi:10.1111/eea.12019.

Restaino, C., Young, D., Estes, B., Gross, S., Wuenschel, A., Meyer, M. and Safford, H. 2019. Forest structure and climate mediate drought-induced tree mortality in forests of the Sierra Nevada, USA. *Ecol. Appl.* 29, e01902.

Rhainds, M., Kettela, E. G. and Silk, P. J. 2012. CP Alexander review 1: thirty-five years of pheromone-based mating disruption studies with *Choristoneura fumiferana* (Clemens) (Lepidoptera: Tortricidae). *Can. Entomol.* 144(3), 379–95. doi:10.4039/tce.2012.18.

Rieske, L. K. 2007. Success of an exotic gallmaker, *Dryocosmus kuriphilus*, on chestnut in the USA: a historical account. *EPPO Bull.* 37(1), 172-4. doi:10.1111/j.1365-2338.2007.01100.x.

Rigsby, C. M., Showalter, D. N., Herms, D. A., Koch, J. L., Bonello, P. and Cipollini, D. 2015. Physiological responses of emerald ash borer larvae to feeding on different ash species reveal putative resistance mechanisms and insect counter-adaptations. *J. Insect Physiol.* 78, 47–54. doi:10.1016/j.jinsphys.2015.05.001.

Rigsby, C. M., Herms, D. A., Bonello, P. and Cipollini, D. 2016. Higher activities of defense-associated enzymes may contribute to greater resistance of Manchurian ash to emerald ash borer than a closely related and susceptible congener. *J. Chem. Ecol.* 42(8), 782–92. doi:10.1007/s10886-016-0736-5.

Robert, L. E., Sturtevant, B. R., Cooke, B. J., James, P. M. A., Fortin, M. J., Townsend, P. A., Wolter, P. T. and Kneeshaw, D. 2018. Landscape host abundance and configuration regulate periodic outbreak behavior in spruce budworm *Choristoneura fumiferana*. *Ecography* 41(9), 1556–71. doi:10.1111/ecog.03553.

Roe, A. D., Torson, A. S., Bilodeau, G., Bilodeau, P., Blackburn, G. S., Cui, M., Cusson, M., Doucet, D., Griess, V. C., Lafond, V., Paradis, G., Porth, I., Prunier, J., Srivastava, V., Tremblay, E., Uzunovic, A., Yemshanov, D. and Hamelin, R. C. 2019. Biosurveillance of forest insects: part I–integration and application of genomic tools to the surveillance of non-native forest insects. *J. Pest. Sci.* 92(1), 51-70. doi:10.1007/s10340-018-1027-4.

Roland, J. 1993. Large-scale forest fragmentation increases the duration of tent caterpillar outbreak. *Oecologia* 93(1), 25–30. doi:10.1007/BF00321186.

Roland, J. 2005. Are the "seeds" of spatial variation in cyclic dynamics apparent in spatially-replicated short time-series? An example from the forest tent caterpillar. *Ann. Zool. Fenn.* 42, 397–407.

Roland, J. and Taylor, P. D. 1997. Insect parasitoid species respond to forest structure at different spatial scales. *Nature* 386(6626), 710–3. doi:10.1038/386710a0.

Rossi, J.-P., Garcia, J., Roques, A. and Rousselet, J. 2016. Trees outside forests in agricultural landscapes: spatial distribution and impact on habitat connectivity for forest organisms. *Landsc. Ecol.* 31(2), 243–54. doi:10.1007/s10980-015-0239-8.

Rudinsky, J. A. and Michael, R. R. 1973. Sound production in Scolytidae: stridulation by female *Dendroctonus* beetles. *J. Insect Physiol.* 19(3), 689–705. doi:10.1016/0022-1910(73)90078-4.

Rudinsky, J. A., Ryker, L. C., Michael, R. R., Libbey, L. M. and Morgan, M. E. 1976. Sound production in Scolytidae: female sonic stimulus of male pheromone release in two *Dendroctonus* beetles. *J. Insect Physiol.* 22(12), 1675–81. doi:10.1016/0022-1910(76)90061-5.

Ryall, K. L., Silk, P. J., Mayo, P., Crook, D., Khrimian, A., Cossé, A. A., Sweeney, J. and Scarr, T. 2012. Attraction of *Agrilus planipennis* (Coleoptera: Buprestidae) to a volatile pheromone: effects of release rate, host volatile, and trap placement. *Environ. Entomol.* 41(3), 648–56. doi:10.1603/EN11312.

Safranyik, L. and Wilson, B. 2006. *The Mountain Pine Beetle: A Synthesis of Biology, Management, and Impacts on Lodgepole Pine*. Natural Resources Canada, Canadian Forest Service, Victoria, BC.

Safranyik, L., Carroll, A. L., Régnière, J., Langor, D. W., Riel, W. G., Shore, T. L., Peter, B., Cooke, B. J., Nealis, V. G. and Taylor, S. W. 2010. Potential for range expansion of mountain pine beetle into the boreal forest of North America. *Can. Entomol.* 142(5), 415–42. doi:10.4039/n08-CPA01.

Sambaraju, K. R., Carroll, A. L., Zhu, J., Stahl, K., Moore, R. D. and Aukema, B. H. 2012. Climate change could alter the distribution of mountain pine beetle outbreaks in western Canada. *Ecography* 35(3), 211–23. doi:10.1111/j.1600-0587.2011.06847.x.

Santos, M. J. and Whitham, T. G. 2010. Predictors of *Ips confusus* outbreaks during a record drought in southwestern USA: implications for monitoring and management. *Environ. Manag.* 45(2), 239–49. doi:10.1007/s00267-009-9413-6.

Schiebe, C., Blaženec, M., Jakuš, R., Unelius, C. R. and Schlyter, F. 2011. Semiochemical diversity diverts bark beetle attacks from Norway spruce edges. *J. Appl. Entomol.* 135(10), 726–37. doi:10.1111/j.1439-0418.2011.01624.x.

Schmidt, B. C., Roland, J. and Wakarchuk, D. 2003. Evaluation of synthetic pheromones for monitoring forest tent caterpillar (Lepidoptera: Lasiocampidae) populations. *Environ. Entomol.* 32(1), 214–9. doi:10.1603/0046-225X-32.1.214.

Schroeder, M. and Dalin, P. 2017. Differences in photoperiod-induced diapause plasticity among different populations of the bark beetle *Ips typographus* and its predator *Thanasimus formicarius*. *Ag. For. Entomol.* 19(2), 146–53. doi:10.1111/afe.12189.

Schwartzberg, E. G., Jamieson, M. A., Raffa, K. F., Reich, P. B., Montgomery, R. A. and Lindroth, R. L. 2014. Simulated climate warming alters phenological synchrony between an outbreak insect herbivore and host trees. *Oecologia* 175(3), 1041–9. doi:10.1007/s00442-014-2960-4.

Scriven, G., Reeves, E. and Luck, R. 1986. Beetle from Australia threatens eucalyptus. *Calif. Ag.* 40, 4–6.

Seidl, R. and Rammer, W. 2017. Climate change amplifies the interactions between wind and bark beetle disturbances in forest landscapes. *Landsc. Ecol.* 32(7), 1485–98. doi:10.1007/s10980-016-0396-4.

Seybold, S. J., Penrose, R. L. and Graves, A. D. 2016. Invasive bark and ambrosia beetles in California Mediterranean forest ecosystems. In: Paine, T. D. and Lieutier, F. (Eds), *Insects and Diseases of Mediterranean Forest Systems*. Springer International Publishing AG, Cham, Switzerland, pp. 583–662.

Seybold, S. J., Bentz, B. J., Fettig, C. J., Lundquist, J. E., Progar, R. A. and Gillette, N. E. 2018. Management of western North American bark beetles with semiochemicals. *Annu. Rev. Entomol.* 63, 407–32. doi:10.1146/annurev-ento-020117-043339.

Seybold, S. J., Klingeman III, W. E., Hishinuma, S. M., Coleman, T. W. and Graves, A. D. 2019. Status and impact of walnut twig beetle in urban forest, orchard, and native forest ecosystems. *J. For.* 117(2), 152–63. doi:10.1093/jofore/fvy081.

Sharov, A. A. and Liebhold, A. M. 1998. Model of slowing the spread of gypsy moth (Lepidoptera: Lymantriidae) with a barrier zone. *Ecol. Appl.* 8(4), 1170–9. doi:10.1890/1051-0761(1998)008[1170:MOSTSO]2.0.CO;2.

Sharov, A. A., Liebhold, A. M. and Roberts, E. A. 1996a. Spatial variation among counts of gypsy moths (Lepidoptera: Lymantriidae) in pheromone-baited traps at expanding population fronts. *Environ. Entomol.* 25(6), 1312–20. doi:10.1093/ee/25.6.1312.

Sharov, A. A., Liebhold, A. M. and Roberts, E. A. 1996b. Spread of gypsy moth (Lepidoptera: Lymantriidae) in the central Appalachians: comparison of population boundaries obtained from male moth capture, egg mass counts, and defoliation records. *Environ. Entomol.* 25(4), 783–92. doi:10.1093/ee/25.4.783.

Sharov, A. A., Liebhold, A. M. and Roberts, A. E. 1998. Optimizing the use of barrier zones to slow the spread of gypsy moth (Lepidoptera: Lymantriidae) populations in North America. *J. Econ. Entomol.* 91(1), 165–74. doi:10.1093/jee/91.1.165.

Sharov, A. A., Leonard, D., Liebhold, A. M. and Clemens, N. S. 2002a. Evaluation of preventative treatments in low-density gypsy moth populations using pheromone traps. *J. Econ. Entomol.* 95, 1205–15.

Sharov, A. A., Leonard, D., Liebhold, A. M., Roberts, E. A. and Dickerson, W. 2002b. "Slow the spread": a national program to contain the gypsy moth. *J. For.* 100,

Shore, T. L. and Safranyik, L. 1992. *Susceptibility and Risk Rating Stands for the Mountain Pine Beetle in Lodgepole Pine Stands*. Natural Resources Canada, Canadian Forest Service, Victoria, B.C.

Showalter, D. N., Raffa, K. F., Sniezko, R. A., Herms, D. A., Liebhold, A. M., Smith, J. A. and Bonello, P. 2018. Strategic development of tree resistance against forest pathogen and insect invasions in defense-free space. *Front. Ecol. Evol.* 6, 124. doi:10.3389/fevo.2018.00124.

Sidder, A. M., Kumar, S., Laituri, M. and Sibold, J. S. 2016. Using spatiotemporal correlative niche models for evaluating the effects of climate change on mountain pine beetle. *Ecosphere* 7(7), e01396. doi:10.1002/ecs2.1396.

Simmons, M. J., Lee, T. D., Ducey, M. J., Elkinton, J. S., Boettner, G. H. and Dodds, K. J. 2014. Effects of invasive winter moth defoliation on tree radial growth in eastern Massachusetts, USA. *Insects* 5(2), 301–18; doi:10.3390/insects5020301.

Six, D. L., Biber, E. and Long, E. 2014. Management for mountain pine beetle outbreak suppression: does relevant science support current policy? *Forests* 5(1), 103–33. doi:10.3390/f5010103.

Smith, G. F. and Nicholas, N. S. 1998. Patterns of overstory composition in the fir and fir-spruce forests of the Great Smoky Mountains after balsam woolly adelgid infestation. *Am. Midl. Nat.* 139(2), 340–52. doi:10.1674/0003-0031(1998)139[0340:POOCIT]2.0.CO;2.

Smitley, D. R., Doccola, J. J. and Cox, D. L. 2010. Multiple-year protection of ash trees from emerald ash borer with a single trunk injection of emamectin benzoate, and single year protection with an imidacloprid basal drench. *Arbor. Urban For.* 36, 206–11.

Sniezko, R. A. and Koch, J. 2017. Breeding trees resistant to insects and diseases: putting theory into application. *Biol. Invas.* 19(11), 3377–400. doi:10.1007/s10530-017-1482-5.

Sobek-Swant, S., Crosthwaite, J. C., Lyons, D. B. and Sinclair, B. J. 2012a. Could phenotypic plasticity limit an invasive species? Incomplete reversibility of mid-winter deacclimation in emerald ash borer. *Biol. Invas.* 14(1), 115–25. doi:10.1007/s10530-011-9988-8.

Sobek-Swant, S., Kluza, D. A., Cuddington, K. and Lyons, D. B. 2012b. Potential distribution of emerald ash borer: what can we learn from ecological niche models using Maxent and GARP? *For. Ecol. Manag.* 281, 23–31. doi:10.1016/j.foreco.2012.06.017.

Stahl, K., Moore, R. and McKendry, I. 2006. Climatology of winter cold spells in relation to mountain pine beetle mortality in British Columbia, Canada. *Clim. Res.* 32, 13–23. doi:10.3354/cr032013.

Stephen, F. M. and Browne, L. E. 2000. Application of Eliminade™ parasitoid food to boles and crowns of pines (Pinaceae) infested with *Dendroctonus frontalis* (Coleoptera: Scolytidae). *Can. Entomol.* 132(6), 983–85. doi:10.4039/Ent132983-6.

Sterner, T. E. and Davidson, A. G. 1982. Forest insect and disease conditions in Canada 1981. FIDS Report. Environment Canada, Canadian Forestry Service, Ottawa, 58pp.

Stewart, D., Zahiri, R., Djoumad, A., Freschi, L., Lamarche, J., Holden, D., Cervantes, S., Ojeda, D. I., Potvin, A., Nisole, A., Béliveau, C., Capron, A., Kimoto, T., Day, B., Yueh, H., Duff, C., Levesque, R. C., Hamelin, R. C. and Cusson, M. 2016. A multi-species TaqMan PCR assay for the identification of Asian gypsy moths (*Lymantria* spp.) and other invasive lymantriines of biosecurity concern to North America. *PLoS ONE* 11(8), e0160878. doi:10.1371/journal.pone.0160878.

Streifel, M. A., Tobin, P. C., Hunt, L., Nadel, H., Molongoski, J. J. and Aukema, B. H. 2017. Landscape-level patterns of elevated FS1 Asian allele frequencies in populations of gypsy moth (Lepidoptera: Erebidae) at a northern U.S. boundary. *Environ. Entomol.* 46(2), 403–12. doi:10.1093/ee/nvx041.

Streifel, M. A., Tobin, P. C., Kees, A. M. and Aukema, B. H. 2019. Range expansion of *Lymantria dispar dispar* (L.)(Lepidoptera: Erebidae) along its north-western margin in North America despite low predicted climatic suitability. *J. Biogeogr.* 46(1), 58–69. doi:10.1111/jbi.13474.

Sun, J., Lu, M., Gillette, N. E. and Wingfield, M. J. 2013. Red turpentine beetle: innocuous native becomes invasive tree killer in China. *Annu. Rev. Entomol.* 58, 293–311. doi:10.1146/annurev-ento-120811-153624.

Susaeta, A., Soto, J. R., Adams, D. C. and Hulcr, J. 2016. Pre-invasion economic assessment of invasive species prevention: a putative ambrosia beetle in Southeastern loblolly pine forests. *J. Environ. Manag.* 183(3), 875–81. doi:10.1016/j.jenvman.2016.09.037.

Swain, K. M. and Remion, M. C. 1981. Direct control of the southern pine beetle. Agric. Handbook 575. United States Department of Agriculture, Forest Service, Pineville, LA.

Szabla, K. 2013. Praktyczna realizacja strategii ograniczania liczebności kornika drukarza na przykładzie świerczyn Beskidu Śląskiego i Żywieckiego. In: Grodzki, W. (Ed.), *Kornik Drukarz i Jego Rola w Ekosystemach Leśnych*. Centrum Informacyjne Lasów Państwowych, Warsaw, pp. 161–75.

Szalay-Marzso, L. 1971. Biology and control of the fall webworm (*Hyphantria cunea* Drury) in the middle-and East European countries. *EPPO Bull.* 1(3), 23–31. doi:10.1111/j.1365-2338.1971.tb02577.x.

Telford, A., Cavers, S., Ennos, R. A. and Cottrell, J. E. 2015. Can we protect forests by harnessing variation in resistance to pests and pathogens? *Forestry* 88(1), 3-12. doi:10.1093/forestry/cpu012.

Thistle, H. W., Peterson, H., Allwine, G., Lamb, B. K., Strand, T., Holsten, E. H. and Shea, P. J. 2004. Surrogate pheromone plumes in three forest trunk spaces: composite statistics and case studies. *For. Sci.* 50, 610-25.

Thomas, V. G., Hanner, R. H. and Borisenko, A. V. 2016. DNA-based identification of invasive alien species in relation to Canadian federal policy and law, and the basis of rapid response management. *Genome* 59(11), 1023-31. doi:10.1139/gen-2016-0022.

Thompson, L. M., Faske, T. M., Banahene, N., Grim, D., Agosta, S. J., Parry, D., Tobin, P. C., Johnson, D. M. and Grayson, K. L. 2017. Variation in growth and developmental responses to supraoptimal temperatures near latitudinal range limits of gypsy moth *Lymantria dispar* (L.), an expanding invasive species. *Physiol. Entomol.* 42(2), 181-90. doi:10.1111/phen.12190.

Thomson, A. J. and Benton, R. 2007. A 90-year sea warming trend explains outbreak patterns of western spruce budworm on Vancouver Island. *For. Chron.* 83(6), 867-9. doi:10.5558/tfc83867-6.

Tisserat, N., Cranshaw, W., Leatherman, D., Utley, C. and Alexander, K. 2009. Black walnut mortality in Colorado caused by the walnut twig beetle and thousand cankers disease. *Plant Health Prog.* 10(1). doi:10.1094/PHP-2009-0811-01-RS.

Tobin, P. C. and Blackburn, L. M. (Eds). 2007. Slow the Spread: a national program to manage the gypsy moth. Gen. Tech. Rep. NRS-6. United States Department of Agriculture, Forest Service, Northern Research Station, Newtown Square, PA, 109pp.

Tobin, P. C. and Hajek, A. E. 2012. Release, establishment, and initial spread of the fungal pathogen *Entomophaga maimaiga* in island populations of *Lymantria dispar*. *Biol. Control* 63(1), 31-9. doi:10.1016/j.biocontrol.2012.06.004.

Tobin, P. C. and Liebhold, A. M. 2011. Gypsy moth. In: Simberloff, D. and Rejmánek, M. (Eds), *Encyclopedia of Biological Invasions*. University of California Press, Berkeley, CA, pp. 298-304.

Tobin, P. C., Sharov, A. A., Liebhold, A. A., Leonard, D. S., Roberts, A. E. and Learn, M. R. 2004. Management of the gypsy moth through a decision algorithm under the STS Project. *Am. Entomol.* 50(4), 200-9. doi:10.1093/ae/50.4.200.

Tobin, P. C., Liebhold, A. M., Roberts, E. A. and Blackburn, L. M. 2015. Estimating spread rates of non-native species: the gypsy moth as a case study. In: Venette, R. C. (Ed.), *Pest Risk Modelling and Mapping for Invasive Alien Species*. Wallingford, UK: CAB International: pp. 131-45.

Unelius, C. R., Schiebe, C., Bohman, B., Andersson, M. N. and Schlyter, F. 2014. Non-host volatile blend optimization for forest protection against the European spruce bark beetle, *Ips typographus*. *PLoS ONE* 9(1), e85381. doi:10.1371/journal.pone.0085381.

USDA APHIS. 2015. Field release of the parasitoid *Spathius galinae* for the biological control of the emerald ash borer (*Agrilus planipennis*) in the contiguous United States (document ID: APHIS-2014-0094-014). Available at: https://www.regulations.gov/document?D=APHIS-2014-0094-0014.pdf.

USDA APHIS. 2016. Pest alert: Asian gypsy moth. APHIS 81-35-027. Issued April 2016, 2p (15 January 2018).

USDA APHIS. 2019. Emerald ash borer regulatory information. Available at: https://www.aphis.usda.gov/aphis/ourfocus/planthealth/plant-pest-and-disease-programs/pests-and-diseases/emerald-ash-borer/ct_regulatory.

USDA Forest Service and APHIS. 1993. Russia and United States exotic pest monitoring program, 2pp. Available at: https://www.fs.usda.gov/Internet/FSE_DOCUMENTS/fsbdev2_025710.pdf (accessed on 15 January 2018).

Valenta, V., Moser, D., Kapeller, S. and Essl, F. 2017. A new forest pest in Europe: a review of Emerald ash borer (*Agrilus planipennis*) invasion. *J. Appl. Entomol.* 141(7), 507–26. doi:10.1111/jen.12369.

Van Asch, M., Salis, L., Holleman, L. J. M., Van Lith, B. and Visser, M. E. 2013. Evolutionary response of the egg hatching date of a herbivorous insect under climate change. *Nat. Clim. Change* 3(3), 244–8. doi:10.1038/nclimate1717.

van Frankenhuyzen, K., Lucarotti, C. and Lavallée, R. 2016. Canadian contributions to forest insect pathology and to the use of pathogens in forest pest management. *Can. Entomol.* 148(S1), S210–38. doi:10.4039/tce.2015.20.

van Lierop, P., Lindquist, E., Sathyapala, S. and Franceschini, G. 2015. Global forest area disturbance from fire, insect pests, diseases and severe weather events. *For. Ecol. Manag.* 352, 78–88. doi:10.1016/j.foreco.2015.06.010.

Vanhanen, H., Veteli, T. O., Paivinen, S., Kellomaki, S. and Niemela, P. 2007. Climate change and range shifts in two insect defoliators: gypsy moth and nun moth-a model study. *Silva Fenn.* 41(4), 621. doi:10.14214/sf.469.

Venette, R. C., Kriticos, D. J., Magarey, R. D., Koch, F. H., Baker, R. H. A., Worner, S. P., Gómez Raboteaux, N. N., McKenney, D. W., Dobesberger, E. J., Yemshanov, D., De Barro, P. J., Hutchison, W. D., Fowler, G., Kalaris, T. M. and Pedlar, J. 2010. Pest risk maps for invasive alien species: a roadmap for improvement. *BioScience* 60(5), 349–62. doi:10.1525/bio.2010.60.5.5.

Veteli, T. O., Lahtinen, A., Repo, T., Niemelä, P. and Varama, M. 2005. Geographic variation in winter freezing susceptibility in the eggs of the European pine sawfly (*Neodiprion sertifer*). *Ag. For. Entomol.* 7(2), 115–20. doi:10.1111/j.1461-9555.2005.00259.x.

Villari, C., Herms, D. A., Whitehill, J. G. A., Cipollini, D. and Bonello, P. 2016. Progress and gaps in understanding mechanisms of ash tree resistance to emerald ash borer, a model for wood-boring insects that kill angiosperms. *New Phytol.* 209(1), 63–79. doi:10.1111/nph.13604.

Villari, C., Sniezko, R., Rodriguez-Saona, L. and Bonello, P. 2017. Accelerating dynamic genetic conservation efforts: use of FT-IR spectroscopy for the rapid identification of trees resistant to destructive pathogens. In: Sniezko, R. A., Man, G., Hipkins, V., Woeste, K., Gwaze, D., Kliejunas, J. T. and McTeague, B. A. (Eds), *Gene Conservation of Tree Species–Banking on the Future*. Proceedings of a workshop. Gen. Tech. Rep. PNW-GTR-963. US Department of Agriculture, Forest Service, Pacific Northwest Research Station, Portland, OR, p. 69.

Virtanen, T., Neuvonen, S., Niemelä, P., Nikula, A. and Varama, M. 1996. Climate change and the risks of *Neodiprion sertifer* outbreaks on Scots pine. *Silva Fenn.* 30(2–3), 169–77. doi:10.14214/sf.a9229.

Volkovitsh, M. G. and Karpun, N. N. 2017. A new invasive species of buprestid beetles in the Russian fauna: Lamprodila (Palmar) festiva (L.) (Coleoptera, Buprestidae), a pest of Cupressaceae. *Entomol. Rev.* 97(4), 425–37. doi:10.1134/S0013873817040042.

Vose, J. M., Wear, D. N., Mayfield III, A. E. and Nelson, C. D. 2013. Hemlock woolly adelgid in the southern Appalachians: control strategies, ecological impacts, and potential management responses. *For. Ecol. Manag.* 291, 209–19. doi:10.1016/j.foreco.2012.11.002.

Vose, J. M., Peterson, D. L., Domke, G. M., Fettig, C. J., Joyce, L. A., Keane, R. E., Luce, C. H., Prestemon, J. P., Band, L. E., Clark, J. S., Cooley, N. E., D'Amato, A. and Halofsky, J. E. 2018. Forests. In: Reidmiller, D. R., Avery, C. W., Easterling, D. R., Kunkel, K. E., Lewis, K. L. M., Maycock, T. K. and Stewart, B. C. (Eds), *Impacts, Risks, and Adaptation in the United States: Fourth National Climate Assessment* (vol. II). U.S. Global Change Research Program, Washington DC, pp. 223–58. Chapter 6.

Wallander, E. 2008. Systematics of *Fraxinus* (Oleaceae) and evolution of dioecy. *Plant System. Evol.* 273(1-2), 25–49. doi:10.1007/s00606-008-0005-3.

Ward, S. F., Venette, R. C. and Aukema, B. H. 2019. Cold tolerance of the invasive larch casebearer and implications for invasion success. *Ag. For. Entomol.* 21(1), 88–98. doi:10.1111/afe.12311.

Watson, G. W., Voegtlin, D. J., Murphy, S. T. and Foottit, R. G. 1999. Biogeography of the *Cinara cupressi* complex (Hemiptera: Aphididae) on Cupressaceae, with description of a pest species introduced into Africa. *Bull. Entomol. Res.* 89(3), 271–83. doi:10.1017/S0007485399000395.

Weed, A. S., Ayres, M. P. and Hicke, J. A. 2013. Consequences of climate change for biotic disturbances in North American forests. *Ecol. Monogr.* 83(4), 441–70. doi:10.1890/13-0160.1.

Weed, A. S., Bentz, B. J., Ayres, M. P. and Holmes, T. P. 2015. Geographically variable response of *Dendroctonus ponderosae* to winter warming in the western United States. *Landsc. Ecol.* 30(6), 1075–93. doi:10.1007/s10980-015-0170-z.

Wermelinger, B. 2004. Ecology and management of the spruce bark beetle *Ips typographus* - a review of recent research. *For. Ecol. Manag.* 202(1-3), 67–82. doi:10.1016/j.foreco.2004.07.018.

Whitehead, R. J., Safranyik, L. and Shore, T. L. 2006. Preventive management. In: Safranyik, L. and Wilson, B. (Eds), *The Mountain Pine Beetle - A Synthesis of Biology, Management, and Impacts on Lodgepole Pine*. Natural Resources Canada, Canadian Forest Service, Victoria, BC, pp. 173–92.

Whitehill, J. G. A., Yuen, M. M. S., Henderson, H., Madilao, L., Kshatriya, K., Bryan, J., Jaquish, B. and Bohlmann, J. 2019. Functions of stone cells and oleoresin terpenes in the conifer defense syndrome. *New Phytol.* 221(3), 1503–17. doi:10.1111/nph.15477.

Witzgall, P., Kirsch, P. and Cork, A. 2010. Sex pheromones and their impact on pest management. *J. Chem. Ecol.* 36(1), 80–100. doi:10.1007/s10886-009-9737-y.

Wood, S. L. and Bright Jr., D. E. 1992. *A Catalog of Scolytidae and Platypodidae (Coleoptera), Part 2: Taxonomic Index. Volume 13, Great Basin Naturalist Memoirs.* Brigham Young University, Provo, UT, 1553pp.

Wulder, M. A., White, J. C., Ortlepp, S. M., Mora, B., Coggins, S., Coops, N. C. and Heath, J. 2012. Digital high spatial resolution aerial imagery to support forest health monitoring: the mountain pine beetle context. *J. Appl. Remote Sens.* 61, 062527. doi:10.1117/1.JRS.6.062527.

Yanchuk, A. D., Murphy, J. C. and Wallin, K. F. 2008. Evaluation of genetic variation of attack and resistance in lodgepole pine in the early stages of a mountain pine beetle outbreak. *Tree Genet. Genomes* 4(2), 171–80. doi:10.1007/s11295-007-0098-9.

Yang, Z. Q., Wang, X. Y., Wei, J. R., Qu, H. R. and Qiao, X. R. 2008. Survey of the native insect natural enemies of *Hyphantria cunea* (Drury) (Lepidoptera: Arctiidae) in China. *Bull. Entomol. Res.* 98(3), 293–302. doi:10.1017/S0007485308005609.

Yang, Z.-Q., Wand, X.-Y. and Zhang, Y.-N. 2014. Recent advances in biological control of important native and invasive forest pests in China. *Bio. Control* 68, 117–28. doi:10.1016/j.biocontrol.2013.06.010.

Yasyukevich, V. V., Titkina, C. N., Davidovich, E. A. and Yasyukevich, N. V. 2015. Changes in range boundaries of the gypsy moth and the nun moth (*Lymantria dispar*, *L. monacha*, *Lymantriidae*, *Lepidoptera*) due to the global warming: a model approach. *Entomol. Rev.* 95(8), 1144–8. doi:10.1134/S0013873815080229.

Zas, R., Björklund, N., Sampedro, L., Hellqvist, C., Karlsson, B., Jansson, S. and Nordlander, G. 2017. Genetic variation in resistance of Norway spruce seedlings to damage by the pine weevil *Hylobius abietis*. *Tree Genet. Genomes* 13(5), 111. doi:10.1007/s11295-017-1193-1.

Zhang, Q.-H. and Schlyter, F. 2004. Olfactory recognition and behavioural avoidance of angiosperm nonhost volatiles by conifer-inhabiting bark beetles. *Agric. For. Entomol.* 6(1), 1–20. doi:10.1111/j.1461-9555.2004.00202.x.

Zheng, X. L., Li, J., Yang, Z. D., Xian, Z. H., Wei, J. G., Lei, C. L., Wang, X. P. and Lu, W. 2014. A review of invasive biology, prevalence and management of *Leptocybe invasa* Fisher & La Salle (Hymenoptera: Eulophidae: Tetrastichinae). *Afr. Entomol.* 22(1), 68–79. doi:10.4001/003.022.0133.

Chapter 4

Using *Verticillium albo-atrum* WCS850 to control Dutch elm disease

Joeke Postma, Wageningen University & Research, The Netherlands

1 Introduction

Trees are crucial for quality of life in urban areas. They limit the urban heat island effect by reducing temperature, improving air quality, contributing to water infiltration, reducing noise disturbance, supporting biodiversity and, above all, improving well-being. Given the importance of these ecosystem services in urban areas, as well as the fact that a single tree can absorb up to 150 kg of CO_2 per year, the 'Trees in Cities Challenge' was launched at the UN Climate Action Summit 2019 to increase the number of trees in cities (https://treesincities.unece.org/).

However, new pests and diseases have appeared in several tree species in recent decades. Examples of pests and diseases in trees that are commonly used in urban areas are: horse chestnut leaf miner (*Cameraria ohridella*), red palm weevil (*Rhynchophorus ferrugineus*), ash dieback (*Chalara fraxinea*) and bleeding canker in chestnut (*Pseudomonas syringae* pv. *aesculi*). Caterpillars of the oak processionary moth (*Thaumetopoea processionea*) are an increasingly serious problem since, in addition to damaging trees they also cause skin irritation and asthma in humans and animals. Given potential risks to human health, chemical pesticides to control pests and diseases in urban areas should be avoided. Biological control is an environmental-friendly alternative (Cazorla and Mercado-Blanco 2016). However, the development of effective

http://dx.doi.org/10.19103/AS.2021.0093.17

biocontrol agents for trees is more complicated due to their perennial nature than for annual crops (Cazorla and Mercado-Blanco 2016). However, 29 years of experience with biological control of Dutch elm disease using the fungus *Verticillium albo-atrum* illustrates the potential for biocontrol of tree pests and diseases. The socio-economic value of a single tree makes individual injection affordable, whereas such treatment is not realistic for individual arable crop plants. Removal of infected trees and transplanting new trees costs 1000 to 3000 Euros per tree depending on their size and the location (Veilleux et al. 2012; personal communication Ron Schraven, BTL Bomendienst). This chapter describes this successful example of biocontrol using *Verticillium albo-atrum*.

A century ago the devastating Dutch elm disease (DED) emerged in Europe (1919) and in North America (1930). Sudden wilting and dying of leaves and branches were caused by the *Ophiostoma ulmi* fungus during this first pandemic from the 1920s to 1940s. An even more aggressive strain of the *O. novo-ulmi* fungus started spreading after 1972 and is responsible for the current pandemic affecting elms. This strain has also infected large numbers of newly planted trees that were supposed to be less susceptible to DED. Since the 1920s DED has killed tens of thousands of elms annually in Europe and North America.

Transmission of the disease is due to a very effective infection cycle with bark beetles as the vector. Bark beetles of the genus *Scolytus* reproduce in weakened and dead (infected) elms. Beetles emerging from infected trees transport conidia of the fungus to healthy elms during their feeding activity on elm twigs (Scheffer et al. 2008). The fungus subsequently develops in the wood vessels, and the infected and weakened tree becomes suitable for breeding of beetles under the stem bark. Large and small elm beetles, *Scolytus scolytus* and *S. multistriatus*, respectively, are the most common and important species spreading the pathogen globally (Santini and Faccoli 2015). Other elm bark beetles are also mentioned as vectors (Anderbrant and Schlyter 1987).

Different methods to control DED have been applied with varying success since the first outbreak of the disease. Chemical control of beetles is potentially hazardous to people and harmful to the environment whilst fungicide (macro)injection in trees is laborious and injures trees (Costonis 1981). In the Netherlands, a national programme begun in the 1930s controlled the disease initially but became unaffordable after the devastating emergence of the even more aggressive strain *O. novo-ulmi*. Strict sanitation measures in some areas such as Amsterdam have reduced incidence to approximately 1% of the elms, including resistant cultivars (Hiemstra et al. 2006). The efficacy of monitoring and strict sanitation was recently evaluated in Canada, showing that rapid removal of infected trees before August reduced the infection rate to 1.5% (Veilleux et al. 2012). New elm cultivars with a high level of resistance to DED are now available on the market (Buiteveld et al. 2015; Martín et al. 2015). This is an important solution for the future but does not prevent existing trees from infection.

A biological control method for DED was developed in the Netherlands in 1980. A strain of *Verticillium albo-atrum* protecting elm trees from DED has been commercially available in several countries since 1992. The biocontrol strain, its mode of action, the product, its application, efficacy and limitations in controlling DED are described in the following sections.

2 Development of the *Verticillium albo-atrum* WCS850 strain

Research on biological control as an alternative for chemical treatments started in the 1980s. It was found that resistance towards *O. novo-ulmi* could be induced in elms that were only resistant to the less-aggressive *O. ulmi* strains by inoculating a mixture of a pathogenic and non-pathogenic strain (Scheffer et al. 1980). This promising result of reducing symptoms by inoculating a non-pathogenic strain began a process of testing antagonistic pseudomonads to control the disease (Scheffer et al. 2008; Scheffer 1983). It was found that the in vitro antagonistic activity of different bacterial isolates did not correlate with the reduction of DED symptoms after injection in trees (Scheffer 1983). Moreover, the effect of bacterial inoculation was dependent on the elm tree clone or cultivar (Scheffer 1989). Both findings supported the hypothesis that induction of resistance in the host was the underlying mechanism rather than antagonism. Unfortunately, these positive biological control effects with bacterial inoculants were only detected in the cultivar 'Commelin' and not in the other clones or cultivars tested.

Subsequent research focused on inducing resistance through different fungi. A non-aggressive *O. ulmi* strain as well as two *Verticillium* strains were tested on the basis that these vascular wilt pathogens could survive within the tree for sufficient time to elicit a resistance response (Scheffer 1990). The isolate *Verticillium albo-atrum* WCS850 (previously identified as *V. dahlia*) proved to suppress disease development effectively in 'Commelin' elms as well as in susceptible field elms (Scheffer et al. 2008). This strain was originally isolated from a potato field in the Netherlands, and is virulent on potato and tomato, but not on elm and several other tree species (Scheffer et al. 2008). It is a naturally hyaline form of *V. albo-atrum* which does not produce resting structures. Conidiospores of the fungus injected in the xylem tissue of trees induce resistance when germinating in the tree.

Efficacy of the *Verticillium albo-atrum* WCS850 strain has been tested in elm trees artificially infected with an aggressive strain of *O. ulmi*. The biocontrol strain was injected at a height of 1.2–1.4 m, and the pathogen was inoculated one or few weeks later at a height of 8 m to mimic the natural infection process. Different clones or species were tested from 1986 to 1988. Over 300 trees aged 15 years old were used in the first stage. Evaluation of disease expression was carried out after 3 months or in the following year. Spores survive during the season but remain close to the site of inoculation (Scheffer 1990). Because elms form a new layer of sapwood each year, the treatment must be repeated every spring.

3 Production, method of application, registration and use

Conidiospores of the isolate *Verticillium albo-atrum* WCS850 are produced under sterile conditions on solid media and are harvested in distilled water, obtaining a suspension with 10^7 spores per mL. Spore concentration, colony formation and absence of impurities and contaminants of each batch are checked. The spore suspensions are stored in customised plastic 40-mL vials at 4°C for a maximum period of 7 weeks.

Trees are injected (vaccinated) with the spore suspension once a year in spring, as soon as the leaves have started to sprout after flowering of the elm, and preferably before any beetle infection of DED occurs. The vaccine is injected into the tree trunk at a height of 1.3 m at every 10 cm of the tree's circumference, by pushing a chisel through the bark and releasing one drop with one pull of the trigger (Fig. 1). The closed injection system prevents spills and any impact on operators, other people, plants, trees or animals in the environment. The method is fast, taking on average 2–3 minutes per treated tree, which enables treatment of large numbers of trees in a short time frame. Only trees without symptoms are vaccinated, since the mode of action of

Figure 1 Hand-held tree injection system: the chisel is pushed through the bark; by pulling the trigger a drop of the vaccine is released which subsequently will be taken up by the elm trunk (photo by BTL Bomendienst).

the biocontrol agent is through induced resistance. A preventative treatment protects the trees against the development of the disease during one season.

All treated trees are checked for DED symptoms in summer and at the end of the year to evaluate the efficacy of the application. A distinction is made between *Ophiostoma* infection starting locally in the tree canopy and spreading downwards, which is due to beetles, and infection starting in the root system which spreads through the whole tree. A typical first sign of an infection transmitted by beetles is an upper branch of the tree with leaves starting to wither and yellow in summer, months before the normal autumnal leaf shedding. The treatment protects against infection starting at the tree canopy, but not against infection through the root system by root grafts.

The commercial name of the product is DutchTrig® with the active substance *Verticillium albo-atrum* WCS850. Although the first field experiments injecting *V. albo-atrum* WCS850 started in 1987 (Scheffer 1990), large-scale application was only initiated from 1992 when the product was registered in the Netherlands (Table 1). After extensive testing in the United States from 1995, full registration by the Environmental Protection Agency (EPA) was granted in 2005. Since 2008 the active substance has been approved for inclusion in the Annex I list of EU Directive 91/414/EEC. Germany, Sweden and Canada were the next list of countries where DutchTrig® was registered, in 2008, 2010 and 2009, respectively. In 2019, the active substance was re-evaluated by the EU under Regulation (EC) No 1107/2009 and approved as a low-risk substance (https://ec.europa.eu/food/plant/pesticides/eu-pesticides-database/public/?event=activesubstance.detail&language=EN&selectedID=2008). The development and placing on the market of low-risk substances and products is encouraged by the EU. Low-risk substances are now approved for 15 years instead of 10 years, which means that the active substance of DutchTrig® is now registered as an active substance in EU until 2034.

Table 1 DutchTrig® registration and application in different countries

	First practical application	Year of registration	Number of treated trees in 2020
The Netherlands	1992	1992	22 000
The United States	1995	2005	3000
Germany	2006	2008	300
Canada	2008	2009	2500
Sweden	2008	2010	1000
UK	2016	2016	0[1]
Norway	2018	2018	700

[1] No applications due to the COVID-19 outbreak.

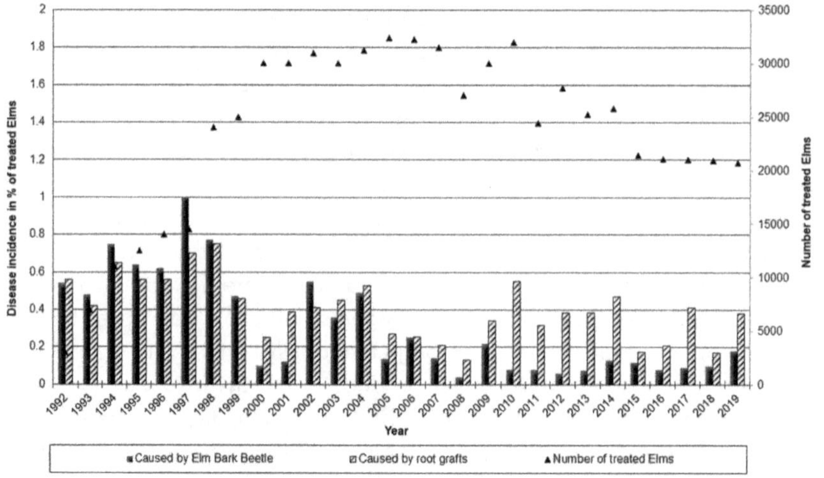

Figure 2 The number of treated elms and disease incidence among the treated trees caused by beetle infections or by root grafts in the Netherlands per year since 1992.

The Netherlands has had the largest number of elms treated with DutchTrig® where 30 000 elm trees were vaccinated annually between 2000 and 2010 (Fig. 2). The number of treated trees subsequently decreased but, since 2015, 22 000 elms have been vaccinated annually. The total number of treated trees in various countries was approximately 29 500 in 2020 (Table 1).

The costs of treatment with DutchTrig® depend on the size and number of trees to be treated. In the Netherlands, where over 20 000 trees are treated annually, the price per tree varies between 16 € and 25 €. These costs are relatively low compared to the removal of dead trees and subsequent transplantation and initial maintenance of new trees. These costs depend on the age and the size of the tree to be removed and transplanted, but are between 1000 and 3000 Euros per tree (Veilleux et al. 2012; personal communication Ron Schraven, BTL Bomendienst).

4 Testing long-term efficacy

As noted, efficacy of the biocontrol strain was first tested in elm trees artificially infected with an aggressive strain of *O. ulmi*. These experiments with 15-year-old elm trees in 1987 and 1988 demonstrated a 70–100% reduction in infected trees (Scheffer 1990). Further efficacy tests with several elm clones were performed in the United States in 2-year-old trees (Scheffer et al. 2008). The most relevant efficacy data are gained from consistently monitoring all treated trees for DED symptoms in the months after treatment (until the end of the year).

In the first years of DutchTrig® application in the Netherlands, from 1992 to 1999, approximately 0.6% of injected trees became infected with DED through beetle transmission (Fig. 2). In this period the number of treated trees increased from 3000 up to 24 000 per year. Between 2000 and 2009 efficacy ranged from 0.1% to 0.6% infection through beetle transmission. However, infection by root grafts cannot be controlled by injection with DutchTrig® and the number of trees infected via this route is, even after 2010, still 0.4% of treated trees (Fig. 2). The infection data of the last decade clearly demonstrate that the treated trees have very low infection levels by beetles, but that the infection through root grafts is often at least twice as high.

To fully assess the efficacy of the biocontrol agent, results should be compared with untreated trees. In the field experiments of 1987 and 1988 treated and untreated trees were tested in combination with artificial pathogen inoculation. In the first years of application in the Netherlands, the natural losses in untreated trees were estimated to range from 4% up to 14% per year approximately, compared to 0.6% infection through beetle transmission in the treated elm trees (Scheffer et al. 2008). However, national figures were approximate since there was no consistent monitoring of all elm trees. More precise efficacy data can be obtained from smaller areas where treated as well as untreated elms are monitored.

In The Hague (the Netherlands) where 9000–10 000 elm trees were injected annually from 1998 to 2001, the infection rate of injected trees decreased from 1% to 0.2%, whereas infection in untreated trees decreased from 5.3% to 3.5%, possibly due to the injection programme reducing disease spread (Scheffer et al. 2008). Similar results are available for Denver (USA) where between 1998 and 2003, approximately 500 out of 1300 elms were injected, resulting in a decrease in DED infection in injected trees from 7% to 0.5% (Scheffer et al. 2008). The infection of untreated trees decreased from 5% to 1%. This can be explained by a decreased number of infected trees suitable for breeding of beetles due to long-term application of the biocontrol agent. More recently, 0.86% of 3416 treated elms in Rotterdam (the Netherlands) became infected between 2007and 2013 compared to 3.47% of 647 untreated elms (Postma and Goossen-van de Geijn 2016).

5 Conclusion

Approximately 100 years after the first introduction of DED in Europe, an effective commercial biocontrol product is available as a preventive treatment to protect susceptible elm trees against DED transmission by beetles. Injection of trees with conidiospores of the fungus *V. albo-atrum* isolate WCS850, product name DutchTrig®, reduces the infection to less than 0.2% of treated trees. This product has been commercially available since 1992 and is now used in seven countries. Other biocontrol agents that have been tested for efficacy against DED are *Verticillium dahliae* isolate Vd-48 (Solla and Gil 2003) and four endophytic fungi (Díaz et al. 2013). However, none of them appeared to be effective. Selecting an effective, environmentally safe and commercially valuable biocontrol agent remains challenging (Köhl et al. 2011).

Practical limitations of the product DutchTrig® are the relatively high production costs and limited shelf-life. Living conidiospores that can germinate in the tree are needed to induce resistance and subsequently control the development of DED symptoms. The short survival period of conidiospores limits shelf life of the product and storage and transport must be at low temperature to keep the spores alive. However, this product has become a commercially viable biocontrol agent, registered and used in Europe as well as in the United States and Canada. An important advantage is that the closed injection system prevents spills, making it safe for the operatives, the public and the environment. An important limitation of DutchTrig® is that it does not protect trees already infected or trees infected via root grafts. Consequently, it is very important to avoid contact of roots between healthy and infected trees. In this respect, it is rather disadvantageous that roads are often planted with one type of tree; mixed plantings would reduce the spread of DED through root contact.

Despite its limitations, DutchTrig® is a valuable component of an integrated control strategy for DED. It can be combined with the removal of trees with DED symptoms to reduce disease pressure, together with the use of new elm cultivars with a high level of resistance to DED (Buiteveld et al. 2015; Martín et al. 2015). In the Netherlands, elm trees are now being replanted in several municipalities. This already had a positive impact on biodiversity illustrated by higher numbers of species such as the white-letter hairstreak butterfly (*Satyrium w-album*). This butterfly needs elm trees for its life cycle and has recently been observed in several Dutch cities by the Dutch butterfly foundation (https://www.vlinderstichting.nl/vlinders/overzicht-vlinders/details-vlinder/iepenpage).

Although the closed injection system is very targeted and safe with no exposure to the environment and applicator, injection of biocontrol agents is not a current practice. Most biocontrol agents are applied by spraying on plants or mixing through soil or substrate before sowing or planting the crop. Nevertheless, for perennial crops injecting biocontrol agents into the stem (or trunk) or vascular system is a viable alternative. Injection with DutchTrig® has been tested against few other pathogens in other tree species, but until now this has not resulted in new potential biocontrol applications (personal communication Ron Schraven, BTL Bomendienst). Other examples of injection are seedling vaccination by stem injecting a conidial suspension of a non-pathogenic *Fusarium oxysporum* strain to suppress *Verticillium* wilt of eggplant (Gizi et al. 2011) and injection of *Trichoderma* spp. and *Paecilomyces lilacinus* into the pseudostem of banana to control the nematode *Radopholus similis*

(Vargas et al. 2015). An example of a commercially available product used to vaccinate tomatoes is V10® containing two non-aggressive strains of PepMV (VX1 and VC1) giving cross-protection against Pepino mosaic virus (VALTO, de Lier, the Netherlands). Here the vaccination is done by rubbing the product into one leaf per plant. Although injection of plants with biocontrol agents is rarely applied, it is particularly of interest for perennial or larger sized crops (trees, grape, kiwifruit, hop, banana, tomato crop under greenhouse cultivation) when internalised pests and diseases are an issue (Rabiey et al. 2019), or when the biocontrol acts through systemic induced resistance.

6 Where to look for further information

Practical information about the product and its application, as well as details per country, can be found at: https://dutchtrig.com/.

7 Acknowledgements

We thank BTL Bomendienst for their support and for supplying data on the number of treated and DED-infected elm trees.

8 References

Anderbrant, O. and Schlyter, F. (1987). Ecology of the Dutch elm disease vectors *Scolytus laevis* and *S. Scolytus* (Coleoptera: Scolytidae) in southern Sweden. *Journal of Applied Ecology* 24(2), 539–550.

Buiteveld, J., Van Der Werf, B. and Hiemstra, J. A. (2015). Comparison of commercial elm cultivars and promising unreleased Dutch clones for resistance to *Ophiostoma novo-ulmi*. *iForest–Biogeosciences and Forestry* 8(2), 158–164.

Cazorla, F. M. and Mercado-Blanco, J. (2016). Biological control of tree and woody plant diseases: an impossible task? *BioControl* 61(3), 233–242.

Costonis, A. C. (1981). Tree infection: perspective macro-injection/micro-injection. *Journal of Arboriculture* 7(10), 275–277.

Díaz, G., Córcoles, A. I., Asencio, A. D. and Torres, M. P. (2013). In vitro antagonism of *Trichoderma* and naturally occurring fungi from elms against *Ophiostoma novo-ulmi*. *Forest Pathology* 43(1), 51–58.

Gizi, D., Stringlis, I. A., Tjamos, S. E. and Paplomatas, E. J. (2011). Seedling vaccination by stem injecting a conidial suspension of F2, a non-pathogenic Fusarium oxysporum strain, suppresses Verticillium wilt of eggplant. *Biological Control* 58(3), 387–392.

Hiemstra, J. A., Buiteveld, J., Kopinga, J., Kranenborg, K. G., Ravesloot, M. B. M., van der Sluis, B. J. and de Vries, S. M. G. (2006). Belang en toekomst van de iep in Nederland. *Praktijkonderzoek Plant & Omgeving, PPO* 421, 51 pp.

Köhl, J., Postma, J., Nicot, P., Ruocco, M. and Blum, B. (2011). Stepwise screening of microorganisms for commercial use in biological control of plant-pathogenic fungi and bacteria. *Biological Control* 57(1), 1–12.

Martín, J. A., Solla, A., Venturas, M., Collada, C., Domínguez, J., Miranda, E., Fuentes, P., Burón, M., Iglesias, S. and Gil, L. (2015). Seven *Ulmus minor* clones tolerant to *Ophiostoma novo-ulmi* registered as forest reproductive material in Spain. *iForest–Biogeosciences and Forestry* 8(2), 172–180.

Postma, J. and Goossen-van de Geijn, H. (2016). Twenty-four years of Dutch Trig® application to control Dutch elm disease. *BioControl* 61(3), 305–312.

Rabiey, M., Hailey, L. E., Roy, S. R., Grenz, K., Al-Zadjali, M. A. S., Barrett, G. A. and Jackson, R. W. (2019). Endophytes vs tree pathogens and pests: can they be used as biological control agents to improve tree health? *European Journal of Plant Pathology* 155(3), 711–729.

Santini, A. and Faccoli, M. (2015). Dutch elm disease and elm bark beetles: a century of association. *iForest–Biogeosciences and Forestry* 8(2), 126–134.

Scheffer, R. J. (1983). Biological control of Dutch elm disease by *Pseudomonas* species. *Annals of Applied Biology* 103(1), 21–30.

Scheffer, R. J. (1989). Pseudomonas for biological control of Dutch elm disease. III. Field trials at various locations in the Netherlands. *Netherlands Journal of Plant Pathology* 95(6), 305–318.

Scheffer, R. J. (1990). Mechanisms involved in biological control of Dutch elm disease. *Journal of Phytopathology* 130(4), 265–276.

Scheffer, R. J., Heybroek, H. M. and Elgersma, D. M. (1980). Symptom expression in elms after inoculation with combination of an aggressive and a non-aggressive strain of *Ophiostoma ulmi*. *Netherlands Journal of Plant Pathology* 86(6), 315–317.

Scheffer, R. J., Voeten, J. G. W. F. and Guries, R. P. (2008). Biological control of Dutch elm disease. *Plant Disease* 92(2), 192–200.

Solla, A. and Gil, L. (2003). Evaluating *Verticillium dahliae* for biological control of *Ophiostoma novo-ulmi* in *Ulmus minor*. *Plant Pathology* 52(5), 579–585.

Vargas, R., Wang, A., Obregon, M. and Araya, M. (2015). Effect of *Trichoderma* spp., *Paecilomyces* lilacinus and nematicide injection into the pseudostem, on *Radopholus similis* control and banana production. *Agronomia Costarricense* 39(2), 61–76.

Veilleux, J., Leferink, J. and Holliday, N. J. (2012). Rapid removal of symptomatic trees reduces Dutch elm disease infection rates. *Arboriculture and Urban Forestry* 38(3), 99–104.